# Praise for METAHUMAN

"In his brilliant new book, Deepak Chopra successfully argues that consciousness is the sole creator of self, mind, brain, body, and the universe, as we know it. Deepak then teaches us that truly grasping this revolutionary idea will effectively remove the limiting belief systems and negativity that may be holding us back from achieving our maximum human potential. Highly recommended!"

> —DR. RUDOLPH E. TANZI, Professor of Neurology,
> Harvard Medical School; and bestselling coauthor of
> *The Healing Self, Super Brain,* and *Super Genes*

"*Metahuman* is a brilliant vision of human potential and how we can move beyond the limitations, concepts, and stories created by the mind. If we wake up and drop our usual habits of considering ourselves as finite, localized beings, the potential exists in each one of us to be one with our very true nature, or quoting Huxley, to be one with the Mind at Large."

> —DR. MENAS C. KAFATOS, *New York Times* bestselling author
> and Fletcher Jones Endowed Professor of Computational
> Physics, Chapman University

"Deepak Chopra's wonderful book *Metahuman* points the way for all of us to realize our full potential by showing us how to shed the constraints that hold us back from a life filled with love and self-worth. I applaud how he roots his recommendations in the latest research. As always, Dr. Chopra has written a book filled with insight, an indispensable aid for anyone striving to have a happy and fulfilled existence."

> —LEONARD MLODINOW, bestselling author of *Elastic:*
> *Unlocking Your Brain's Ability to Embrace Change*

"This book distills decades of personal practice and breadth of study, and boils it all down to this: we are more than we think, more even than the universe itself, because the universe is only one expression of the fundamental, grounding, metahuman source that emanates, nurtures, and transcends us all. The book reads as though an old, wise friend, away traveling for years, collecting adventures and stories, comes home."

> —NEIL THEISE, M.D., Professor of Pathology, New York
> University School of Medicine

"*Metahuman* is a powerful wake-up call! Brilliantly merging scientific and holistic insights, Deepak Chopra allows us to go beyond illusory mental constructs to witness reality without limits. Only then can we reach our full potential!"
>           —DR. LARS BUTTLER, cofounder and CEO of the
>           Artificial Intelligence Foundation

"Deepak Chopra reveals a fresh way of approaching special states of consciousness and gives us new insights into ancient ideas in the contemplative traditions. Highly recommended."
>           —BERNARD J. BAARS, PH.D., former Senior Fellow in
>           Theoretical Neurobiology, the Neurosciences Institute,
>           San Diego, California

"In *Metahuman*, Deepak Chopra exposes us to a new and exciting reality, where we are all infinitely free to create the lives we yearn for. There is no better guide to lead us to this phenomenal way of being."
>           —BRUCE VAUGHN, CEO, Dreamscape Immersive; and former
>           Chief Creative Executive at Walt Disney Imagineering

"For over thirty years, Deepak Chopra has been educating us on the extraordinary potential of the human being, focusing initially on our capacity for exceptional physical and mental health, and then for spiritual wellbeing. *Metahuman* takes us beyond health and wellbeing—and all such concepts of the mind—to a life based on our very own nature and that of the universe: existence itself."
>           —PAUL J. MILLS, PH.D., Professor and Chief, Family
>           Medicine and Public Health; Director, Center of Excellence
>           for Research and Training in Integrative Health, University of
>           California, San Diego

"Reading *Metahuman* is not just *about* waking up, it is *to wake up*. It is the GPS for realizing your full potential. You need it, I need it, and we all need it if we are to live on this planet without destroying it and ourselves. The key is in us—it *is* us. And Deepak Chopra tells us how to reach it."
>           —ERVIN LASZLO, author of *Science and the Akashic Field* and
>           *Reconnecting to the Source*

"Deepak Chopra and I have had our differences on a few scientific issues, but the one thing that I've always admired about him is how he applies the methods and findings of science to social progress, especially to help people improve their lives. We all want to lead happy and meaningful lives, but how? *Metahuman* is Deepak's answer to this deepest of all personal questions. I learned something new in every chapter toward this important end. You will be enlightened by the research and wisdom he brings together."

—MICHAEL SHERMER, publisher of *Skeptic* magazine; Presidential Fellow, Chapman University; and author of *Heavens on Earth*

"In public health, we look for ways to make the healthy choice the easy choice. *Metahuman* does this by mapping a direct path to a state of focused awareness and to self-regulation mechanisms, based on growing evidence from mindfulness research at Brown. This has the power to help those suffering from stress lower their blood pressure as well as reduce their risk for anxiety, depression—even loneliness."

—BESS H. MARCUS, PH.D., Dean, Brown University School of Public Health

"Drawing upon examples ranging from medical science to parable, Deepak Chopra illustrates how, like virtual reality technologies, the human mind paints a layer of imagination atop of the physical world."

—D. FOX HARRELL, PH.D., Professor, Digital Media and Artificial Intelligence; and Director, MIT Center for Advanced Virtuality, Massachusetts Institute of Technology

"Reflecting on neuroscience, physics, cosmology, and anthropology— everything from the molecular-level sensitivity of our senses to an octopus's emotions—Deepak Chopra explains how the world of our experience is an inner construct and not the thing in itself. In so doing, he offers a path through modern life, guiding us to put suffering in perspective and perhaps gain some control over it."

—GEORGE MUSSER, author of *Spooky Action at a Distance* and *The Complete Idiot's Guide to String Theory*

"It takes courage to bridge the gap between hard-core 'western' science and eastern esoteric quasi-mystical traditions. There is no one better equipped to construct such a bridge than the erudite Deepak Chopra, who transmits the 'wisdom of the east' to his audience. The result is this highly readable book that will stimulate readers' curiosity about human existence and the nature of reality."

—V. S. RAMACHANDRAN, Professor of Neuroscience, University of California, San Diego

"This book contains the quintessence of Deepak Chopra's teachings. Everything comes forth from a conscious source; our own consciousness can link us back to it, if only we cease to separate ourselves from it. Chopra's argument is fluent, scientifically well-informed, and persuasive, as well as eminently practical, with detailed instructions. How refreshing!"
—RUPERT SHELDRAKE, PH.D., biologist and author of *Ways to Go Beyond and Why They Work*

"*Metahuman* is a reflective deconstruction of the everyday experience of reality, away from the physical and toward the mental and metaphysical. Here Deepak Chopra aptly explores the evolution of consciousness and charts a path for humans to become liberated and limitless."
—MICHELLE A. WILLIAMS, SC.D., Dean of Faculty, Harvard T. H. Chan School of Public Health

"In *Metahuman,* Deepak Chopra gives us a beatifically written understanding of his views on consciousness, the universe, the body, and the mind. This book is a tremendous resource—not only for providing a direct path to attain true reality but, importantly, for enabling a healthier, more wholesome life."
—RALPH SNYDERMAN, M.D., Chancellor Emeritus, Duke University; and Director, Duke Center for Personalized Health Care

"This book is a manual for living as nature intended us to, and it can't be read soon enough. Deepak Chopra's clarity and gentleness, and his ability to reach people—despite the maelstrom of nonsense and confusion we find ourselves immersed in today—probably make him one of the most important people alive."
—BERNARDO KASTRUP, PH.D., author of *The Idea of the World*

"Deepak Chopra's prose is a work of art, an inspired vision of the life of the mind and what it means to be human and beyond. Science will continue to explore the true origins of consciousness, but until we have definitive answers, this book provides practical tools for increasing human well-being. Even reductionists like myself will find much to enjoy in Deepak's latest contribution to the philosophy of mind."
—HEATHER BERLIN, PH.D., M.P.H., Assistant Clinical Professor, Department of Psychiatry, Icahn School of Medicine at Mount Sinai

"Integrating recent discoveries in neuroscience, endocrinology, and psychology, Deepak Chopra demonstrates in concrete terms how reductionistic and physicalistic constraints on human nature are simply incorrect and seriously limit those who hold them. This book is a page-turner."
—ALLAN LESLIE COMBS, PH.D., Professor of Consciousness Studies, California Institute of Integral Studies

"Deepak Chopra's *Metahuman* is a deep dive into educating or familiarizing its readers with a fundamental understanding of self, consciousness, enlightenment, and *metareality*, in the same way Stephen Hawkins's *A Brief History of Time* attempted to teach the layman about the mysteries of the origins of the physical quantum universe. A must-read."
—DR. KENNETH P. GREEN, D.M.D., B.S., Commander, US Navy (retired); Senior Vice President, Brain Mapping Foundation; Vice President, Strategic Initiatives for Government and Nonprofit Partnerships; and Member of the Board of Directors, Society for Brain Mapping and Therapeutics

"In *Metahuman*, Deepak Chopra provides us with a step-by-step guide for how we can once again reimagine the reality of the world we live in, a reality not bound by the beliefs and understandings we have about the material and physical world, but by the boundlessness of our purpose and energy."
—ROBERT LoCASCIO, founder and CEO, LivePerson

"What sets Deepak Chopra's new book apart and makes it particularly valuable and important is that it takes modern life and secular culture as its point of departure, rather than being situated in the context of the spiritual wisdom tradition of some ancient and distant culture. *Metahuman* does not shy away from what those approaches offer but, importantly, is also not bound by their perspectives and approaches. This book, Dr. Chopra's gentle and insightful guidance, is a great blessing for its readers."
—EDWIN L. TURNER, Professor, Department of Astrophysical Sciences, Princeton University

"This book is another masterpiece from the renowned endocrinologist Deepak Chopra. It is an invitation to self-discovery, self-awareness, and self-improvement. A must-read for those who seek to be better humans, live much more fulfilling lives, and impact others."
—BABAK KATEB, M.D., Chairman of the Board and CEO, Society for Brain Mapping and Therapeutics

# METAHUMAN

## UNLEASHING YOUR INFINITE POTENTIAL

### Deepak Chopra, M.D.

HARMONY BOOKS

Published in the United States by Harmony Books,
an imprint of Random House, a division of
Penguin Random House LLC, New York.
harmonybooks.com

Harmony Books is a registered trademark, and the Circle
colophon is a trademark of Penguin Random House LLC.

Library of Congress Cataloging-in-Publication Data
Names: Chopra, Deepak, author.
Title: Metahuman: unleashing your infinite potential / by
Deepak Chopra.
Description: New York: Harmony, [2019] |
Includes bibliographical references and index. |
Identifiers: LCCN 2019004165 (print) |
LCCN 2019006154 (ebook) | ISBN 9781524762988
(e-book) | ISBN 9780307338334 (hardcover: alk. paper)
Subjects: LCSH: Mind and body. | Spiritual life. |
Self-realization.
Classification: LCC BF161 (ebook) | LCC BF161 .C458
2019 (print) | DDC 204/.4—dc23
LC record available at https://lccn.loc.gov/2019004165

ISBN 978-0-307-33833-4
Ebook ISBN 978-1-5247-6298-8
International Edition ISBN 978-0-593-13609-6

Printed in the United States of America

Book design Meighan Cavanaugh
Jacket design by Pete Garceau

1  3  5  7  9  10  8  6  4  2

FIRST EDITION

# CONTENTS

.........................................

A Personal Preface: Going Beyond   *1*

Overview: Metahuman Is the Choice of a Lifetime   *9*

## PART ONE

## THE SECRETS OF METAREALITY

1. We Are Entangled in an Illusion   *35*

2. "I" Is the Creator of Illusion   *59*

3. Human Potential Is Infinite   *77*

4. Metareality Offers Absolute Freedom   *97*

5. Mind, Body, Brain, and Universe Are Modified Consciousness   *115*

6. Existence and Consciousness Are the Same   *137*

PART TWO

# WAKING UP

7. Putting Experience First    159

8. Going Beyond All Stories    177

9. The Direct Path    193

PART THREE

# BEING METAHUMAN

10. Freeing Your Body    215

11. Recovering the Whole Mind    231

12. Choiceless Awareness    249

13. One Life    263

# A MONTH OF AWAKENING: 31 METAHUMAN LESSONS    277

A Final Word    337

Acknowledgments    343
Index    345

# GOING BEYOND

This book is an invitation to find out who you really are, beginning with two simple questions. In moments when you feel very happy, do you also watch yourself being happy? When you happen to get angry, is some part of you totally free of anger? If you answer "yes" to both questions, you can stop reading. You have arrived. You have gone beyond everyday awareness, and this going beyond is what it takes to know who you really are. Self-knowledge will unfold for you every day. In time—or perhaps at this very moment—you will see yourself living in the light. Like the great Bengali poet Rabindranath Tagore, you can say, "That I exist is a perpetual surprise."

It would be fascinating to meet you, because your existence no doubt is quite unusual—you might even assume that you are unique. You look around and see that the vast majority of people are simply happy when they are happy, and angry when they are angry. But not you. You see beyond.

When I began writing books thirty years ago, there was no question that being happy and getting angry were normal, without the added element of watching yourself. A word like *mindfulness* wasn't in the air;

meditation was still considered dubious by the average person, and the whole question of higher consciousness was viewed with hard-eyed skepticism. I was a young Boston doctor with a growing family, and my days were consumed with work, servicing a large patient roster and traveling every day between two or more hospitals.

When I was happy about a patient with a thyroid condition getting well, did I watch myself being happy? Absolutely not. If the wrong prescription was filled by a careless pharmacist, was a part of me not at all upset, standing by like a silent witness? No. In common with everyone else I knew, I was happy or angry without any mystery about it. But coming from India, I could reach back into my childhood for clues about a different state of being. According to an ancient Upanishad, the human mind is like two birds sitting on a branch. One of the birds is eating the fruit of the tree while the other lovingly looks on.

Since I went for some years to a school run by an order of Catholic brothers, there were other clues from a different source, such as Jesus telling his disciples to be "in the world but not of it." If you Google that phrase, you'll find a wealth of confusion about what it actually means, but the kernel of the teaching is that there is a difference between buying into worldly life and not buying into it. When you don't buy into it, Jesus teaches, you are somehow with God.

I wish I could say that these clues about higher consciousness transfixed me and shaped my life. They didn't. I stored them in the back of my mind, never calling upon them in my busy, stress-filled life. There was no budding awareness of the Truth with a capital T, which is that I, and everyone else in the world, embody the mystery of existence. This is the reason, ultimately, why Tagore found himself perpetually surprised. Once you wake up to reality, you face the mystery of existence intimately and personally: there could be no mystery without you.

In a sentence or two I've taken some giant leaps, I know. There's a yawning gulf between the things a person must do in a day—beginning with waking up, getting dressed, going to work, and so on—and the mystery of existence. A society based on reason and science looks with skepticism on any such notion as being in the world but not of it, or Truth with a capital T. We live together in a reality that obeys the rule of "What you see is what you get." The physical world confronts us; we come to grips with its many challenges; and, as the rational mind probes the dark unknown, what emerges are new facts and data, not a sense of wonder that we even exist.

What first coaxed me into facing the mystery of life—and the mystery of myself as a human being—was medicine. I practiced endocrinology, a specialty that fascinated me because hormones are unique chemicals. They can make you sluggish and dull if you have a thyroid deficiency; they can make you run away or fight when confronted by a threat. A burst of adrenaline is responsible for a common reaction to a street magician levitating before our very eyes as onlookers jump back or run away.

We are so used to accepting that these behaviors are chemically induced that almost everyone connects adolescent behavior with "raging hormones." Even when sexual drive is tamed somewhat, it is never truly tamed, just as falling in love is never rational. If I had been satisfied to accept this commonsense connection between hormones and the effects they cause, there would be no more story.

But there is a fly in the ointment, and it disrupts things far beyond hormones—it potentially overturns reality itself. There is a brain hormone called oxytocin that has gained the popular name of the "love hormone," because the presence of higher levels of this hormone in the brain makes a person more affectionate and trusting. But this one molecule secreted by the pituitary gland is much more complex than that. Higher levels are secreted in the mother during birth and breastfeeding, promoting a close bond with the baby. If you pet your dog for a while, oxytocin goes up in both you and your dog. Oxytocin makes people love their national flag more, while being indifferent to the flags of other countries.

During sexual activity, oxytocin rises in women, making them bond with their sexual partners emotionally, but the effect doesn't seem to happen in men.

Something strange must be going on, and yet these complex findings don't shake the faith of most endocrinologists. I was different. What bothered me was that oxytocin doesn't actually do anything it is credited with doing unless the mind goes along with it. A woman won't have more affection for a sexual partner if she is coerced, frightened, angry, or simply distracted by something more important. Your oxytocin won't go up if you pet a dog you dislike. You won't love your country's flag if you are forced to salute it by an authoritarian regime.

I came to see the explosive effect of the mind-body connection. It was as if we are two creatures, one a robot that can be programmed by chemicals, the other a free agent who thinks, considers, and decides. These two creatures are seemingly incompatible. They have no right to exist together, and yet they do, as reflected in the setup of our nervous system. One part operates automatically, enabling life to go on without your thinking about it. You don't have to think to keep breathing or have your heart beat. But you can consciously take control, and the voluntary nervous system allows you to alter your breathing and even, with a little practice, slow down your heart rate.

Suddenly, we are on the verge of a mystery, because *something* must decide whether to take action, or not. That something cannot be the brain, because the brain is indifferent about whether it employs either side of the central nervous system. On the involuntary side, the brain increases your heart rate if you run a marathon, but it was *you* who decided to run the marathon in the first place.

So who is this "you"?

That niggling question is what disrupts reality. At any given moment you—that is, the self—decide which nervous system to call upon; therefore, you cannot be the creation of either one. When you see this simple fact, you are on the road to self-awareness. You can be happy and watch yourself being happy at the same time; you start to experience yourself completely without anger, even as you are displaying anger.

The reason for this shift is simple: you have gone beyond the mechanical side of life. You have awakened to who you really are, the user of the brain but not the brain, the traveler in a body but not the body, the thinker of thoughts who is far, far more than any thought. As I will show in the following pages, your true self is beyond time and space. When you identify with your true self, you have fulfilled the dictum to be in the world but not of it. The Greek word *meta* means "beyond," so I'm using it to describe the reality that lies beyond "What you see is what you get." When you occupy metareality, you are metahuman.

In fits and starts, everyone is already there. Metareality is the source of all creativity, because without going beyond the old and conventional, there would be no new thoughts, artworks, books, or scientific discoveries. No matter how many thoughts you've had in your life, there are infinitely more you can think; no matter how many sentences writers have written, there are infinitely more to write. Words and thoughts are not stored in the brain like information in a computer, to be juggled around mechanically when another thought is needed. Shakespeare wasn't simply juggling his Elizabethan vocabulary—he employed words in a creative way. Van Gogh didn't simply combine the standard colors in the spectrum; he used color as a new way of seeing the world around him.

Going beyond is how a person decides if life is meaningful enough. When you want more than your life is giving you, it's not your brain that craves more meaning, nor is it the everyday person going about the routine business of life. The self, viewing things from a higher perspective, is deciding the matter. The self also decides whom to love, what is truth, whether to trust, and so on. If a mother judges that a cranky three-year-old needs a nap, she has gone beyond a simple assessment of what the child is doing and saying. Cranky children say all kinds of things, and if mothers bought into them, they'd be no better than children.

If going beyond has proven so indispensable, why aren't we

metahuman already? There is no reason to keep repeating the same trite, tired opinions, follow the same outworn social conventions, and surrender to conformist thinking. All pose traps that we fall into, and the result is more of the same strife, wars, domestic violence, racial prejudice, and gender inequality that we have been prey to throughout history. We choose to be our own prisoners. This paradox, playing the part of inmate and jailer at the same time, has caused untold suffering for humanity.

To bring the whole sorry mess to an end involves one thing: shifting from human to metahuman. Both states exist here and now. There is nowhere to go to reach metareality. Like the two birds in the tree, you are feasting on life while also looking on. But the looking-on part is being ignored, suppressed, overlooked, and undervalued. The transformation that makes you metahuman is known in the world's spiritual traditions as "waking up." Once someone rises to the state of metahuman, it seems as if the old everyday self was a sleepwalker, barely conscious of life's infinite possibilities.

To be awake is to embrace full self-awareness. Lots of other metaphors come to mind. Metahuman is like tuning in to the whole radio band instead of one narrow channel. It's like a string vibrating to a higher note. It's like seeing a world in a grain of sand. But *like* is a limiting word. The real thing is indescribable and must be experienced firsthand, just as sight is indescribable to someone born blind and yet revelatory if that person gains sight.

Editors encourage writers to coax readers, using a big promise of something, new, fresh, and different. Waking up is as old as being human. It's impossible to promise something like waking up, which is indescribable in the first place. Looking back on my previous works, I feel that I was inhibited by how peculiar and mysterious it is to wake up. This time, however, I've taken a deep breath and gone for broke. I'm trusting that the reader isn't someone born blind to whom sight is completely unknowable. With a modicum of trust, we can all be shown that we are already metahuman and that metareality is here and now.

I don't know who will be persuaded and who won't. In the end, t... mystery of being human obeys only itself. But one thing I do have fait... in. If in reading this book you connect with what it means to wake up, you will realize the truth in much less time than the thirty years I look back on. The faster that metahuman dawns in our lives, the better.

# METAHUMAN IS THE
# CHOICE OF A LIFETIME

There are many things people do to improve their lives. You might say that developed societies live in a golden age, as far as standard of living goes. It has become realistic to look forward to decades of good health, to eat organic whole foods available around the corner, not to mention having things that were once out of reach for the average person, such as owning your own home and retiring in relative security.

It is strange, then, that millions of people strive to improve their lives without improving their personal reality. The two are intimately entwined, and if you don't improve your reality, there's something shaky and unreliable about improving your life. Reality isn't simply the world "out there"—it is very personal. Two commuters going to the same job might look at the world entirely differently, one feeling anxious about job security and the prospect of being fired, the other placidly content and optimistic. Giving birth could be the same physical event, without any medical complications, for two new mothers, but one might suffer from postpartum depression while the other is filled with maternal joy.

Personal reality defines us. It consists of all the things we believe in, the emotions we feel, our unique set of memories, and a lifetime of experiences and relationships. Nothing is more decisive in how a person's life turns out. So it is peculiar—one might say profoundly mysterious—that we build our lives on a deep lack of knowledge about who we really are. Delve into any basic issue about human existence, and behind the façade of expert opinion lies a blankness where understanding should be.

We have no idea why humans are designed to both love and hate, preach peace and practice violence, swing between happiness and despair, and lead lives governed by confidence one moment and self-doubt the next. Right now you are acting out in your own fashion all these contradictions. You are a mystery to yourself, because everyone is a mystery to themselves. What keeps people moving forward is the routine of everyday life and the hope that nothing goes horribly wrong.

I'm not devaluing the things most people live for—family, work, and relationships. But, to be blunt, we don't manage even the most important things with any confidence that we know what we're doing. It's no wonder that we spend so much time working to improve our lives and so little working to improve our reality. Reality is too confusing. We are better off ignoring the deep water and remaining where it is safe in the shallows.

A handful of people, however, have ventured into deeper waters, and in every culture they bring back reports that are alien and inspiring at the same time. It's inspiring to love your enemies, but who really does? Being told that divine love is infinite doesn't make it so in your reality. Eternal peace vies with the prospect of crime, war, and violence in every age. A handful of people are cherished as saints, with a good chance they will be labeled as mad instead, or simply dismissed as too good for this world.

Yet one thing is beyond doubt—personal reality is where the whole game is played. It contains all the potential that humans have fulfilled, but also all the limitations that hold us back. A New York psychologist named Abraham Maslow, who died in 1970, continues to be famous today because he swam against the tide. Where the typical career in psychology consisted of examining the ills and defects of the psyche,

Maslow felt that human nature went far beyond everyday experience. His core idea, which has now blossomed far beyond anything he could have imagined, is that humans are designed for extraordinary heights of experience, and, more than that, we should be creating these experiences in everyday life. It was as if the only cars that were on the road were junky rust heaps, and someone announced that you could trade in your clunker for a Mercedes or Jaguar.

If the only cars you see are junk heaps and the Mercedes and Jaguars exist far across the ocean, your reality won't change. But Maslow, drawing on centuries of spiritual aspirations, insisted that the peak experiences in life are part of our design, that we need and crave them. The key was to go beyond the everyday.

The notion of going beyond became the motivation for this book.

To discover who you really are, you must go beyond who you think you are. To find peace, you must go beyond fear. To experience unconditional love, you must go beyond conditional love, the kind that comes and goes. I even thought for a time that this book should simply be titled *Beyond*. Instead I chose *Metahuman*, using the Greek word *meta*, which I noted earlier means "beyond." My thesis is that becoming metahuman is a major shift of identity that anyone can make. Being designed for peak experiences raises the question of whether we have a choice. Often the most illuminating moments in life descend as if from another, higher plane by themselves. How do we know they aren't accidental?

At a recent conference on science and consciousness, a young woman introduced herself, telling me that she was writing her graduate thesis on communicating with birds. I asked her how talking to birds was possible, and she replied that it was easier to show me than to tell me. We went outside. It was a bright day, and we sat quietly on a bench. She looked up at some birds sitting in a tree nearby, and one of them flew down and landed unafraid in her lap.

How did she do it? Feeling no need for words, she gave me a look that said, "See? It's very simple." My old Catholic schoolteachers would have pointed to St. Francis of Assisi, who is often portrayed beatifically with birds fluttering to him. From the Indian tradition, I thought of a quality in consciousness known as *ahimsa*, which means "harmlessness," the empathy extended to all living things.

In either case, it wasn't a matter of talking to the birds or knowing their language—the whole thing had taken place silently. It was a perfect example of going beyond—in this case, going beyond my own expectations. What the young woman did, she explained later, was to have mental clarity and insert an intention for the bird to come to her. In other words, it all happened in consciousness.

So few people have such experiences that it only magnifies the need to show how much choice we really have to go beyond. My strong feeling is that we have much more control over life than we currently realize.

To me, metahuman is the choice of a lifetime. Peak experiences are only the beginning, a glimpse at what is possible.

The term *peak experience* has become popular enough that most people have a general sense of what it means. The term describes moments when limitations drop away and life-changing insights come our way or a superb performance happens effortlessly. The quarterback in NFL football who approaches age forty with multiple Super Bowl wins, the musical prodigy who debuts in a Mozart piano concerto at age eight, the mathematical whiz who can multiply two eighteen-digit numbers in a matter of seconds—we don't have to search far to find stories of peak performance like these that hint at enormously expanded human potential. But these accomplishments, astonishing as they are, occupy a specific niche. When fame and fortune are lavished on the exceptional few, we miss a much greater possibility that applies to the many.

Reality is much more malleable than anyone supposes. Most of the limitations that you feel are imposed on you personally are actually self-imposed. Not knowing who you really are keeps you stuck in secondhand beliefs, nursing old wounds, following outworn conditioning, and suffering a sense of self-doubt and self-judgment. No one's life is free of these limitations. The ordinary world, and our ordinary lives in the world, are not sufficient to reveal who we really are—quite the opposite. The ordinary world has deceived us, and this deception runs so deep that we have molded ourselves to conform to it. In law, tainted evidence is known as the "fruit of the poisonous tree." It's not an exaggeration to say that as good as life can get, there is still a taint stemming from the deceptions we mistake for reality. Nothing, however beautiful and good, has completely escaped this taint. Going beyond is the only way to escape it.

A metahuman is someone whose personality is based on higher values; not just peak experiences, but love and self-worth. After finishing this book I was delighted to find that Maslow had actually used the term *metahuman* in exactly this way. (He didn't associate it with comic book superheroes, and neither do I. While fantasy metahumans are persecuted as freaks and threats to society, this is a connotation to be avoided completely.)

It's all well and good to consider certain experiences so exalted that they seem divine, which is where Maslow placed metahuman. It was an important step to declare that aspiring to reach God or eternal peace and love is just as real as driving in a nail. But I will argue that becoming metahuman is an urgent necessity. It is the only way out of the illusions that play out in our lives as inner suffering, confusion, and conflict.

# The Fantasy of Everyday Life

Everyone would agree that it is better to live in reality than in fantasy. So it will come as a shock that you have been living in a fantasy all your life. It's an all-embracing illusion you bought into from earliest childhood. Even the most practical, hardheaded person is immersed in fantasy all the time. I don't mean flights of fancy or erotic fantasies or dreams of getting rich overnight. Nothing you perceive is as it seems. Everything is an illusion from the ground up.

Take out your smartphone and look at any photo you've saved on it. The image is several inches across, whether the photo is of the Grand Canyon, a mouse, or a microbe. Your eyes are about as far apart as the screen of a smartphone, but you perceive the Grand Canyon, a mouse, and a microbe as hugely different in size. How do we automatically adjust the size of what appears on a smartphone? No one knows, and this becomes even more puzzling when you consider that the retina at the back of the eye is curved and the image projected on it is upside down. Why doesn't the world look as distorted as in a funhouse mirror?

You could shrug your shoulders and ascribe the whole mystery to the brain, which massages the raw data reaching the eye and gives us a realistic picture of the world. But this only deepens the illusion. When we say that our eyes respond to "visible light," we conveniently skip over the fact that the elementary particles of light—photons—are invisible. A photon has no radiance, luster, color, or any other characteristic we associate with light. Like a Geiger counter that clicks madly in the presence of high levels of radioactivity and emits only a few clicks at low levels, the retina "clicks" madly when millions of photons trigger the rods and cones that line it and clicks faintly when light levels are low (which we call darkness).

Either way, everything you think you see has been processed inside your brain, in a specific region known as the visual cortex, which is totally dark. A flashbulb blinding you in the eye is just as black in the brain as the faintest glimmer of stars at night. Nor do the signals reaching the visual cortex form pictures, much less 3-D images. The picture you take to be a snapshot of the world was fabricated by your mind.

In the same fashion, the other four senses are just "clicks" on the surface of other kinds of cells. There is no explanation for why the nerve endings in your nose should turn the bombardment of molecules floating around into the scent of a rose or the stink of a garbage dump. The entire three-dimensional world is based on a magic trick no one can explain, but it is certainly not a true picture of reality. The whole thing is mind-made.

A neuroscientist would stop and correct me, claiming that the world we perceive is brain-made instead. A few simple examples disprove this contention, however. As far as your brain is concerned, the letters on this page are black specks, no different from specks you might scatter randomly with flecks of ink off a brush. Before you learned to read the alphabet, letters were only meaningless specks, while after you learned to read they became meaningful. Yet you had the same brain from three years old onward, as far as processing information goes. The mind learns to read, not the brain. Likewise, anything you see around you—an elm tree, a Belgian chocolate bar, a church, or a cemetery—acquires meaning because your mind gives it meaning.

Another example: When children who were born blind are given sight through medical means, they are baffled by things we take for granted. A cow in the distance looks the same size to them as a cat close up. Stairs look painted on the wall; their own shadow is a mysterious black patch that insists on following them around. What such children have missed—and need to catch up with—is the learning curve by which we all learned to shape ordinary reality. (So discomfiting is the visible world that newly sighted children and adults often prefer to sit in the dark to regain a sense of comfort.)

The learning curve is necessary to make your way in the world, but you have adapted yourself in strange and unexpected ways. Take perspective. If you are lying in bed and someone touches your shoulder to wake you up, you don't see the person having a very wide body with a small head on top. But take a photo from a position lying in bed and reality is revealed. The person's torso, being level with your eyes, is unnaturally wide, while the head, being farther away, is unnaturally small. Likewise, when you are talking to someone right next to you, his nose is swollen out of proportion, and if you compare it to a photo, his eyes might be bigger than the hand resting in his lap.

We automatically block out how things really look in perspective, and through an act of mind we adjust the data. The data reaching your eye reports that the room you are sitting in has walls that converge closer together at the far end, but you know that the room is square, so you adjust the data accordingly. You know that a nose is smaller than a hand, requiring a similar adjustment.

What causes real shock is that *everything* you perceive is adjusted. Floating molecules in the garden are adjusted into fragrances. Vibrating airwaves are adjusted into sounds you recognize and identify. There is no escaping that we live in a mind-made world. This is both the glory and the peril of being human. Walking the streets of London two hundred years ago, the visionary poet William Blake lamented over what he saw:

> *[I] mark in every face I meet*
> *Marks of weakness, marks of woe.*
>
> *In every cry of every man,*
> *In every infant's cry of fear,*
> *In every voice: in every ban,*
> *The mind-forged manacles I hear*

It's a woeful picture, still being repeated today. Humans have wandered into every kind of suffering and hardship out of a deep-seated belief that we are destined to lead such an existence. There is no alternative until you accept that what the mind has made, it can unmake.

# Welcome to the House of Illusions

While participating in the everyday world, it isn't possible to see beyond the illusion. Going beyond is needed, which is why the shift to metahuman is needed. The only way an illusion can be all-encompassing is if everything about it is deceptive, fooling us about the big things and the little things alike. That's the case here. The human mind has constructed everything to suit itself from the ground up. In a sense, this book was written simply to convince you that your personal reality is totally mind-made, and not just by your mind alone. Having spent a lifetime adapting to the artificial reality you inherited as a child, you have to undertake a journey to discover the difference between reality and illusion.

To anyone who accepts the physical world "out there" as totally real, the notion of a mind-made world seems absurd. It's one thing to be struck by an idea, but quite another to be struck by lightning. The difference is so obvious that you'd distrust anyone who told you that the two events were the same.

But some of the greatest minds have said just that. This is where the real fascination begins. Max Planck, a brilliant German physicist, was a major figure in the quantum revolution; in fact, he coined the term *quantum mechanics*. In a 1931 interview with the *Observer* newspaper in London, Planck said, "I regard consciousness as fundamental. I regard matter as derivative from consciousness. We cannot get behind consciousness. Everything that we talk about, everything that we regard as existing, postulates consciousness."

In other words, *consciousness is fundamental.* If that's true, then roses blooming in an English garden spring from the same source as a painting of a rose. That source is awareness, meaning your awareness. Without consciousness, nothing can be proven to exist. Simply by being conscious, you participate in the mind-made world and help create it every day. The beauty of this understanding is that if creation springs from consciousness, we can reshape reality from its source.

Planck was not alone in his reinterpretation of reality, away from the physical toward the mental. The whole drift of the quantum revolution was to dismantle the commonsense view that the world is first and foremost material, solid, and tangible. Another brilliant quantum pioneer, the German physicist Werner Heisenberg, said, "What we observe is not nature itself, but nature exposed to our method of questioning."

The implications of this statement are astonishing. Gaze out your window, and you might see a tree, a cloud, a swath of grass, or the sky. Plug any of those words into Heisenberg's sentence in place of the word *nature*. You see a tree because you ask to see a tree. You see a mountain, a cloud, the sky for the same reason. As an observer, everything outside your window comes into being through the questions you are asking. You might not be aware of asking questions, but that's only because they were asked so early on. When toddlers spy their first tree, they test to see what it is, basically asking, "Is this hard or soft? Rough or smooth? Tall or short? What are those green things on the branches? Why do they ripple in the breeze?" In this way, by applying human consciousness to everything in the universe, we get answers that fit human consciousness. But we don't get reality. Physics dismantles every quality of a tree—its hardness, height, shape, and color—by revealing that all objects are actually invisible ripples in the quantum field.

If this discussion seems too abstract, it can be brought very close to home. Your body is being created in consciousness right this minute; otherwise, it couldn't exist. Again, Heisenberg can be credited with getting there early: "The atoms or elementary particles themselves are not real; they form a world of potentialities or possibilities." But in the commonsense world, where the body is our shelter, life support system, and personal vehicle for getting around, defending it becomes necessary. It's too disturbing to think of our body as a mental illusion.

# The Anti-Robot Argument

Shifting away from the false assumption that the world is solid and physical runs counter to a trend I find increasingly disturbing. Science persistently tries to prove that human beings are machines, and where this was once just a metaphor for how the body works in all its complex parts, man-as-machine is being taken more and more literally. We are told that the complexity of human emotions can be reduced to rising and falling levels of brain hormones. Brain areas that light up on an fMRI scan supposedly indicate the causation or mechanism behind a person feeling depressed or being prone to criminal behavior and much else. Besides being brain puppets, we are supposed to believe that our genes program us in powerful ways, to the point that "bad" genes doom a person to a host of problems, from schizophrenia to Alzheimer's. Those examples of predisposition are then extended to behaviors and traits like being prone to anxiety and depression.

Metahuman has many implications, but one of the strongest is to rebuff the notion that human beings are primarily mechanisms. Even though science has a wealth of findings about both genes and the brain, that doesn't make that notion any more valid. The general public isn't aware, for example, that only 5 percent of disease-related genetic mutations will definitely cause a particular illness. The other 95 percent of genes raise or lower a person's risk factors and, in complex ways, interact with other genes.

The public is still stuck on a misconception that a single gene, like the so-called "gay gene" or the "selfishness gene," exists and creates an irresistible predisposition. This misconception was obliterated when the human genome was mapped. The current picture of DNA is almost the opposite of the public's misguided image. DNA isn't fixed; it is fluid and dynamic, constantly interacting with the outside world and with your inner thoughts and feelings.

The notion that your genes run your life is ingrained, even among educated people, so it is eye-opening to review a recent experiment published in the December 10, 2018, issue of *Nature: Human Behavior*. Researchers in the psychology department of Stanford University took two groups of participants and tested them for two genes, one associated with higher risk of becoming obese, the other with higher risk of performing badly in physical exercise.

First I'll focus on the obesity gene. The participants ate a meal and afterward were asked how full they felt; in addition, their blood was tested for levels of leptin, the hormone associated with feeling full after a meal. The results were about the same for people genetically prone to obesity and those who weren't. The next week the same group returned and ate the same meal, but with a difference. Half the group was told, completely randomly, that they had the gene that protects someone from risk for obesity while the other group was told they had the higher-risk version of the gene.

To the surprise of researchers, there was an immediate and dramatic effect. Simply by being told that they had the protective gene, subjects showed a blood level of leptin two and a half times higher than before. The group who were told they didn't have the protective gene didn't change from their earlier results. What this indicates is that simply being told about a genetic benefit caused people to exhibit the physiology associated with that gene. What the participants believed to be true overrode their actual genetic predisposition, because in some cases the people who thought they were genetically protected actually weren't.

The same dramatic results occurred in the exercise experiment. People who were told that they had a gene that produced poor results from exercise displayed the cardiovascular and respiratory signs that such a gene is supposed to produce. Even though they didn't have the at-risk gene, merely by being told they did reduced their lung capacity and made them too exhausted to continue running on a treadmill.

In short, the body conforms to mind-made reality. If your physiology produces genetic effects simply by hearing that you have a certain gene, the myth of genes controlling our lives is seriously challenged. It's not that genetic programming is irrelevant (for the full picture, refer to the book *Super Genes,* which I cowrote with Harvard geneticist Rudy Tanzi), but the reality is as complex as human life itself. Genes are among the host of causes and influences that affect us. How strongly they affect any given person is impossible to predict, and in every area of behavior and health there is wide latitude for personal choice.

Given an either/or choice, see yourself as a free agent capable of conscious change, rather than a robotic machine run by genes and brain cells. Life is rarely as simple as either/or, which is true here as well. But despite the public image fostered by popular science articles, it's not true that a human being is a biological puppet. Far closer to the truth is the view that we are conscious agents whose potential for creativity and change is unlimited. We become metahuman by making the life-altering choice to be metahumans.

## At the Metahuman Crossroads

I don't expect you to accept this conclusion—not yet, anyway. The overall picture needs to be sketched before you make up your mind. Without realizing it, we are all embedded in a preformed reality that we began to adapt to in infancy. Everything you perceive at this moment through the five senses—the solid walls of your room, the faint movement of air in your lungs, the brightness of the light streaming in through the window or emitted by a lamp—is a simulation, a construct that engulfs you in a virtual reality.

On the one hand, we are set up—brain, body, and mind—to conform to virtual reality, the result of a collective hoodwinking that has taken many thousands of years to create. This makes things very tricky. A prisoner has an incentive to dig a tunnel to the outside world, because he knows that there's something lying beyond the prison walls. The virtual reality you now experience offers nothing on the other side that you can touch, taste, feel, hear, or smell. But something does lie outside virtual reality, which I'll term *metareality*. Metareality is the workshop where consciousness creates everything. It is our source and origin, a field of pure creative potential. Metareality is not perceived by the five senses, because it has no shape or location. Yet it is totally accessible, and it offers our only means to escape simulated reality.

Once you realize that you are engulfed in a simulation, it dawns on you how infinite the creative power of humans really is. We fashioned our world using not bricks and mortar but one invisible material: consciousness. In a scientific age, this assertion seems incredible, if not absurd. From inside the simulation, creation can be viewed like a movie of the universe unfolding from the big bang onward, along a time line that has taken 13.7 billion years. How can this mind-boggling arena, bounded by time, space, matter, and energy, be essentially fake?

To find out, it will take a personal sense of curiosity and a touch of adventurousness to go beyond conventional wisdom. Consciousness is present in every second of our lives, yet conventional wisdom takes it for granted. This isn't like missing the forest for the trees. It's like living in the forest without seeing any trees at all.

Take the enormously popular book *Homo Deus*, whose overarching theme is the invention of the future. The author, Israeli historian Yuval Noah Harari, wants to offer a new and better starting point for the future. Age-old burdens from the past once seemed inescapable, Harari writes:

> The same three problems preoccupied the people of twentieth-century China, of medieval India and of ancient Egypt. Famine, plague and war were always at the top of the list. . . . Many thinkers and prophets concluded that famine, plague and war must be an integral part of God's cosmic plan or of our imperfect nature.

In a burst of optimism rare among futurists, Harari goes on to write that these problems are essentially solved, even though they persist in pockets around the globe: "[A]t the dawn of the third millennium, humanity wakes up to an amazing realisation. . . . [I]n the last few decades we have managed to rein in famine, plague and war." Eagerly his readers want to accept Harari's vision that "on the cosmic scale of history humankind can lift its eyes up and start looking towards new horizons."

And what are these horizons? In *Homo Deus*, Harari takes the reader on a journey through all the existing problems and a roster of possible solutions that futurists love to explore. Only on page 409 does he arrive at consciousness, and then he touts a future dominated by "techno-religions"—in other words, our evolution is leading toward artificial intelligence and supercomputers that upgrade the raw material of the human brain. Faced with a stupendous intelligence that towers over us, what can we do but worship it?

Harari's vision winds up in the wrong place because it started in the wrong place. Consciousness belongs on page 1, and the future that evolved consciousness can lead to is where humanity should be heading. Every future that has unfolded throughout history has been based on a direction taken by the mind. Artificial intelligence, after all, is just another notch on the belt of human intelligence; therefore, predicting that we will be surpassed by a Frankenstein race of supercomputers is very premature. We need to know our full capacity before taking bets on any future. Until metareality becomes a common experience, being human has not reached its full creative capacity. Settling for a better dream isn't good enough—an upgraded illusion is still an illusion.

# In Your Life

## THE METAHUMAN SURVEY

The best evidence we have for going beyond is that everyday people are already experiencing metareality. One measure of this is a twenty-question survey developed by John Astin and David Butlein, which goes by the awkwardly academic title of Nondual Embodiment Thematic Inventory (NETI). On a scale from 20 to 100, NETI assesses how people rank themselves on qualities long considered spiritual, psychological, or moral. They include metahuman traits we already value highly because they are so meaningful, as well as other traits that make life easier to cope with, such as the following:

Compassion
Resilience
Propensity to surrender
Interest in truth
Lack of defensiveness
Capacity to tolerate cognitive dissonance (i.e., having inconsistent
    thoughts, beliefs, or attitudes)
Tolerance for emotional discomfort
Gratitude
Low anxiety level
Authenticity
Humility

These traits describe human nature liberated from secondhand social norms and conditioning. When you possess these qualities, you are free to reach the metahuman state of awareness.

Take a moment to participate personally. Here's the NETI questionnaire that was used to assess what are often called "nondual experiences," meaning a heightened state of consciousness. You will give yourself a total score from 20 to 100, and we'll proceed from there.

# NETI Questionnaire[*]

Instructions: Please indicate how often the following things occur for you. Circle only one answer (note: scores are reversed for questions 4, 8, 14, and 16):

1. Never
2. Rarely
3. Sometimes
4. Most of the time
5. All of the time

1. An inner contentment that is not contingent or dependent upon circumstances, objects, or the actions of other people.

    1. Never
    2. Rarely
    3. Sometimes
    4. Most of the time
    5. All of the time

2. Accepting (not struggling with) whatever experience I may be having.

    1. Never
    2. Rarely
    3. Sometimes
    4. Most of the time
    5. All of the time

---

[*] Developed by John Astin and David A. Butlein.

3. An interest in clearly seeing the reality or truth about myself, the world, and others, rather than in feeling a particular way.

     1. Never

     2. Rarely

     3. Sometimes

     4. Most of the time

     5. All of the time

4. A sense that I am protecting or defending a self-image or concept I hold about myself.

     5. Never

     4. Rarely

     3. Sometimes

     2. Most of the time

     1. All of the time

5. Deep love and appreciation for everyone and everything I encounter in life.

     1. Never

     2. Rarely

     3. Sometimes

     4. Most of the time

     5. All of the time

6. Understanding that there is ultimately no separation between what I call my "self" and the whole of existence.

     1. Never

     2. Rarely

     3. Sometimes

     4. Most of the time

     5. All of the time

7. Feeling deeply at ease, wherever I am or whatever situation or circumstance I may find myself in.

    1. Never

    2. Rarely

    3. Sometimes

    4. Most of the time

    5. All of the time

8. A sense that my actions in life are motivated by fear or mistrust.

    5. Never

    4. Rarely

    3. Sometimes

    2. Most of the time

    1. All of the time

9. Conscious awareness of my nonseparation from (essential oneness with) a transcendent reality, source, higher power, spirit, god, etc.

    1. Never

    2. Rarely

    3. Sometimes

    4. Most of the time

    5. All of the time

10. Not being personally invested in or attached to my own ideas and concepts.

    1. Never

    2. Rarely

    3. Sometimes

    4. Most of the time

    5. All of the time

11. An unwavering awareness of a stillness/quietness, even in the midst of movement and noise.

    1. Never

    2. Rarely

    3. Sometimes

    4. Most of the time

    5. All of the time

12. Acting without assuming a role or identity based on my own or others' expectations.

    1. Never

    2. Rarely

    3. Sometimes

    4. Most of the time

    5. All of the time

13. A sense of immense freedom and possibility in my moment-to-moment experience.

    1. Never

    2. Rarely

    3. Sometimes

    4. Most of the time

    5. All of the time

14. A desire to be understood by others.

    5. Never

    4. Rarely

    3. Sometimes

    2. Most of the time

    1. All of the time

15.  Concern or discomfort about either the past or the future.

   1. Never

   2. Rarely

   3. Sometimes

   4. Most of the time

   5. All of the time

16.  A sense of fear or anxiety that inhibits my actions.

   5. Never

   4. Rarely

   3. Sometimes

   2. Most of the time

   1. All of the time

17.  A feeling of profound aliveness and vitality.

   1. Never

   2. Rarely

   3. Sometimes

   4. Most of the time

   5. All of the time

18.  Acting without a desire to change anybody or anything.

   1. Never

   2. Rarely

   3. Sometimes

   4. Most of the time

   5. All of the time

19. Feelings of gratitude and/or open curiosity about all experiences.

    1. Never

    2. Rarely

    3. Sometimes

    4. Most of the time

    5. All of the time

20. A sense of the flawlessness and beauty of everything and everyone, just as they are.

    1. Never

    2. Rarely

    3. Sometimes

    4. Most of the time

    5. All of the time

Total score _____

## EVALUATING YOUR SCORE

If you have *never* experienced any traits of metahuman awareness, your score will be 20. If you experience metahuman awareness *all the time*, your score will be 100. Either one would be extremely rare. The average scores are as follows for three specific groups drawn from the therapeutic community:

Psychology graduate students: 52

Psychotherapists: 71

Psychotherapists self-reporting as being established in nondual
    (i.e., metahuman) awareness: 81.6

What does this say about people in everyday life? Most important, can we all develop higher consciousness here and now? To find out, a research team that I was part of conducted a study on short-term awakening. We got 69 volunteers who were healthy adults, ranging in age from 32 to 86 (the average age was just over 59). There were two requirements: that they would largely abstain from alcohol for a week (one drink per day was allowed) and that they had not been on a meditation or yoga retreat within the past twelve months.

The participants were divided randomly into two groups at the Chopra Center in Carlsbad, California, which offers a spa setting. One group was told to spend the next six days relaxing and enjoying the spa experience. The other group underwent an Ayurveda-based mind-body program aimed at improving overall well-being. This included a special diet (primarily vegetarian but also geared to specific body types), massage, meditation, and instructions in leading an Ayurvedic lifestyle. The approach is far-reaching in that it covers emotional and spiritual well-being. We already recognized from years of offering the program, known as Perfect Health, that participants would report afterward that they felt healthier, less stressed, more relaxed, and generally happier.

The specific angle in this new study was to compare the two groups on how they answered the NETI questionnaire before and after the six days were up. The mind-body group showed a significant improvement in their scores, compared with the control group, and the results were sustained on reevaluation one month later.*

In the Chopra Center study, participants started out with above-

---

* Complete research details were published in a peer-reviewed article in the *Journal of Alternative and Complementary Medicine* in December 2017. The coauthors come from a range of institutions, from the University of California–Davis, Harvard Medical School, Duke University, and the Chopra Center for Wellbeing. Very quickly the article became widely cited. Its lengthy title is "Change in Sense of Nondual Awareness and Spiritual Awakening in Response to a Multidimensional Well-Being Program."

average scores, averaging 62, which is 10 points higher than the typical psychology grad student. After being divided into the two groups, the group that went through the Perfect Health regimen had a mean score of 74 (higher than the average psychotherapist), while the group that went through six days of relaxation in a spa setting improved only marginally, with an average score of 68. When evaluated a month later, there was a small increase among the Perfect Health group, from 74 to 76, while the score for the relaxation group remained flat.

Your score is meant to give you a rough idea of where you stand, with the caveat that this was just one small study. It is striking and hopeful to find that a weeklong focus on mind and body increases these experiences, and the road ahead is open to anyone who wants to develop even better training programs.

I'm not suggesting that the Perfect Health approach based on Ayurveda is the last word for reaching higher consciousness. It's the overall implications that matter the most. Metahuman experiences are everywhere, but people differ in how often they have them. Some people are well established in these experiences, which have occurred frequently over the course of a lifetime. Such people may take for granted an experience like feeling blissful energy in their bodies. The same experience would astound someone else if it came out of the blue and was totally new.

The range of consciousness is far greater than a questionnaire can measure, obviously. Still, an enormous question is posed: Why live in limitation when expanded awareness offers such great rewards, such as the sense of peace and understanding that come when you know who you really are and the unlimited creative potential that you were designed to fulfill?

# THE SECRETS
# OF METAREALITY

# 1

## WE ARE ENTANGLED
## IN AN ILLUSION

Somewhere in prehistory *Homo sapiens* crossed over into virtual reality, when a mind-made simulation became essential in our evolutionary path. The exact era will never be known, or the reason, if any, why one species should acquire such powers and know that it had them. No other creature consciously shapes its future. No other species tells stories and convinces itself that they are true. There are many mysteries in our past. Somehow, following whatever tortuous path, we managed to make our simulation so convincing that we got lost in it.

Although this simulation is very convincing, on a daily basis it breaks down. There are times when life goes out of kilter and the world doesn't seem real and substantial anymore. Such experiences occur regularly, either to ourselves or to other people. For example, when there's a sudden death in the family or a catastrophe like a tornado or the house burning down, we may go into shock. With a blank stare we reveal how dislocated our existence suddenly feels, saying things like "This can't be happening. It's unreal" or "Nothing matters anymore."

Normally, this dissociated state will pass, and in time reality feels real again. But some people never return—after a psychotic break, for

example, a percentage of mental patients become chronically schizo-
phrenic and have hallucinations, seeing images or hearing voices for the
rest of their lives. But the feeling of "This can't be happening; it's like a
dream" doesn't have to be triggered by shock. Countless people engage in
personal fantasies of fame, wealth, or some other dream that feels totally
real to them and drives them all their lives. When someone is suddenly
ecstatically happy, for whatever reason, everything can seem surreal, too.

However, the physical world "out there" feels real and substantial a
lot more than 99 percent of the time, which is proof enough, one would
think, that we aren't under some kind of spell. But we are. Ironically,
there's now technology that forces a person to confront what is real and
what isn't. When you don a virtual reality (VR) headset, powered by
artificial intelligence, the simulation you are plunged into is like a wraparound, three-dimensional movie of such vividness that it overwhelms
the senses and causes a dislocation from what we deem as everyday reality. You might find yourself precariously perched on a steel construc-
tion girder in midair with the city street many stories below. Your brain,
fooled by the visual image, triggers the stress response just as if you were
really teetering on the girder. You will feel yourself going off balance in
a panic, even though in the room where you are actually standing, your
feet are firmly on the ground and you are in no danger of plunging to
your death.

The VR illusion is created by visual images, and the same holds true
in everyday life. What you see, you believe in. Such trust is misplaced, as
every grade-schooler learns when told that the sun doesn't actually rise in
the east and set in the west. Yet when quantum physics tells us that matter
isn't what it appears to be, we continue to cling to the sensations of weight
and solidity of hard physical objects as if they were indisputable. Would
a bullet be less dangerous if you saw through the illusion? No. The bullet
and the entire physical world remain intact but with the realization that
they are the end point of a process that begins in consciousness.

Once you grasp this and fully absorb it, your personal reality becomes much more malleable, because you can go to the source and be part of the creative process. Getting untangled from the virtual-reality simulation isn't easy. Our personal experience would have to change drastically, but the beauty of it is that we have the potential for change where before we had none or very little. While you cannot turn bullets into cotton balls, to accept that all of reality "out there" is beyond your ability to change isn't true.

The ground rules of everyday life are much looser than we imagine. Even when a person feels completely immersed in the simulation, there is an escape route. And not just one, but many. This only makes sense. Metareality is more real than any virtual simulation. We should regard glimpses into it as evidence that we can inhabit the meta state all the time. Instead, the entanglements of virtual reality have turned the picture wrong side up. As you read the meta experiences below, you will be tempted to see them as anomalous, freakish, or untrustworthy. Getting real is a process that begins by confronting your misplaced trust in illusions every day.

# "Something Happened"

Let's consider one of the most basic aspects of virtual reality. Hardly anyone would question that being inside the body is normal, natural, and a true experience. But this certainty runs counter to the phenomenon of out-of-body experiences (OBEs), which have been documented in every culture for centuries. The most widely publicized out-of-body experience is "going into the light," as reported by patients who have clinically died during emergency medical procedures, especially from heart attacks.*

---

* The near-death experience is discussed in detail in my book *Life After Death*, giving evidence offered by skeptics as well as researchers who support the validity of "going into the light."

It turns out that expecting to go into the light when we die is misleading, because what happens in near-death experiences is much more individual than anyone thought. The largest study of near-death experiences, which examined 2,060 patients who died under emergency or intensive care, arrived at the conclusion that death isn't a single event—it is a process. There isn't simply one final or definitive event. During this process, there are ways to reverse death. In cases where medical professionals were successful at getting the heart, lungs, and brain to come back to normal functioning, about 40 percent of those who died and came back remember that "something happened" when they were flatlined.

This part of the study, which was titled AWARE and was led by British intensive-care doctor Sam Parnia, seems irrefutable. But very quickly the details of "something happened" become controversial. We have to dive into a few details to see what the issues are. Out of the 2,060 patients who died (the study went from 2008 to 2012 and included 33 researchers in 15 hospitals), 104 were resuscitated. The first point to note is that all had actually died. They were not "near death." Their hearts and lungs had stopped functioning, and within 20 to 30 seconds their brains showed no activity. The decomposition of cells throughout the body actually takes several hours to commence afterward. During the interval between dying and being brought back is when 39 percent reported the memory of being conscious even though their brains had stopped.

Dr. Parnia believes that this is probably just a fraction of those who had such experiences; the rest had their memories erased either by brain inflammation, which occurs for 72 hours after a person is brought back from death, or because of drugs that are administered as part of resuscitation, which also cause memory loss. Of the 101 patients who completed the questionnaire about their experience during death, only 9 percent had an experience compatible with the typical "going into the light" model. The majority of memories were vague and unfocused, sometimes pleasant but sometimes not.

Only 2 percent of those who came back, which means 2 people out of 101, had the experience of full awareness or out-of-body experiences such as looking down from above their bodies watching and listening to the medical team as it was working to revive them. Only one person could accurately narrate what had been happening in the room in such detail that it corresponded to timed events. So what does this one person tell us about dying?

It depends. Skeptics shrug off all such experiences as purely physical, claiming that if we had finer measurements of brain activity, at a very subtle level we'd discover that the brain hadn't actually died. Dr. Parnia accepts that this might be true. His main focus is on how to achieve better results at resuscitation that might bring back a normal person with no organ damage, particularly brain damage after clinical death. But Dr. Parnia's personal conclusion is that a person can be fully conscious without brain function, as this one patient was. He points to the basic disagreement, thousands of years ago, between Aristotle and Plato. Aristotle contended that consciousness was a physical phenomenon, Plato that it was nonphysical, residing in a soul that transcends the body.

The AWARE study didn't confirm either side. Unsurprisingly, skeptics and believers didn't change their position, or their prejudices. One can say that it's a significant step to turn death into a process that can be reversed. It's also significant that awareness during death covers a wide range of experiences, not a one-size-fits-all of going into the light. What I'd like to underscore is that even when you die, you fashion the experience personally. Dr. Parnia found that people's spiritual interpretation of their death experience coincided with their own faith. They interpreted the light as being Christ, if they were Christians, which was different for Hindus and totally nonspiritual for atheists.

What happens when we die, then, is open to interpretation. The only consensus among those who came back was that death is a comfortable process, not to be feared. Having directly experienced that their fear of

death was groundless, these people discovered a different perspective on life. Many if not most concluded that they should lead more selfless lives in service to others.

I think it is useful that the AWARE study validated that "something happens," but why are we trying to settle the issue of consciousness at the most extreme moment when life and death hang in the balance? It's like trying to validate gravity by asking survivors of a plane crash about their experience of falling from the sky.

It is the normal, everyday experience of consciousness that needs to be explained, not the extreme states. I've debated or conversed with many neuroscientists, and none has been able to answer the simplest questions about consciousness. These include the following:

What is a thought?

How does the electrochemical activity in a neuron turn into words, sights, and sounds in our heads?

Why is a person's next thought totally unpredictable?

If someone has a vocabulary of 30,000 words, does this mean that a clump of brain cells knows 30,000 words? If so, in what way are the words being stored? For the word *cat*, is there a place inside a brain cell that holds the letters *c-a-t*?

No one can adequately answer any of these questions.

# The Self-Model

An experience as otherworldly as "going into the light" might be a red herring. It turns out that being "inside" your body is a malleable state; you can enter and leave your body almost at will.

In a fascinating account in *The New Yorker*, titled "As Real As It Gets" by Joshua Rothman (April 2, 2018), the issue of living inside the body is

confronted with unusual clarity. At nineteen Thomas Metzinger, a German university student, fell asleep on a meditation retreat and woke up feeling an itch in his back. According to Rothman:

> He tried to scratch it, but couldn't—his arm seemed paralyzed. He tried to force the arm to move, and, somehow, this shifted him up and out of his body, so that he seemed to be floating above himself. . . . He heard someone else breathing, and, in a panic, looked around for an intruder. Only much later did he realize that the breathing had been his.

This uncanny experience soon ended but left a lasting impression. As it happened, Metzinger became a prominent philosopher of mind, and he set out assiduously to explain OBEs, which are estimated to occur in 8–15 percent of the population, generally at night or after surgery. As real as his own experience felt—it was followed on occasion by other, similar experiences—Metzinger discovered its limitations. He couldn't flick the light switch, for example, or fly out the window to visit his girlfriend.

A surprising explanation began to dawn on Metzinger. He discovered the work of the psychologist Philip Johnson-Laird and his theory of "mental models." Instead of logically assessing the world, Johnson-Laird held, we apply a mental picture and shift from one mental model to another, depending on the situation. "If you want to know whether a rug will go with your sofa," Rothman explains, "you don't deduce the answer—you imagine it, by moving furniture around on a mental stage set."

Metzinger began to wonder if what we call reality isn't merely a stage set arranged and colored in by the mind. This was a key insight, and confirmation came by chance when he was contacted by a Swiss neuroscientist, Olaf Blanke, who had artificially induced OBEs in his patients. Working with a forty-three-year-old woman with epilepsy, Blanke had stimulated a specific area of her brain with mild electrical current, "and she had the experience of floating upward and looking down at her own body." This illusion had many variations that could be deliberately brought on. Rothman's *New Yorker* article explains:

> Stimulating another location in the brain created the impression of a doppelgänger standing across the room; stimulating a third created the "sense of a presence"—the feeling that someone was hovering nearby, just out of sight.

Metzinger found this research hard to interpret, because he was committed as a philosopher to tracing experience back to the mind, rather than taking the usual scientific approach that all mental events are the products of physical activity in the brain. But he eventually reached a conclusion consistent with the notion of "mental models." Something like a radical breakthrough occurred. Rothman continues:

> It isn't just that we live inside a model of the external world, Metzinger wrote. We also live inside models of our own bodies, minds, and selves. These "self-models" don't always reflect reality, and they can be adjusted in illogical ways. They can, for example, portray a self that exists outside the body—an O.B.E.

This is a fruitful way to explain why living "inside" the body feels so convincing—we need it for stability and safety, to feel grounded inside our personal shelter. There are other ways to induce an OBE, such as using the drug ketamine, which has mind-altering properties. VR simulation is perhaps the most effective, however. For example, with a specific VR setup, Metzinger saw his own body standing in front of him with its back to him (this was done by placing a camera behind him and transmitting the image to the VR headset). If someone scratched Metzinger's back, he felt the sensation happening to the body he saw in front of him—an eerie, distorting feeling. By remaining "inside" your body, you can avert such disorientation.

But, at the same time, you are trapped behind the skin of your protective suit. It's not that an OBE is better than the normal way of inhabiting our bodies, but that we have seemingly lost the ability to shift from one mental model to another. This ability can never be entirely lost, however. In a variety of ways—dreaming, fantasizing, denial, willful blindness, and more—we turn our backs on the simulation we have agreed to accept most of the time.

Virtual reality in any guise—drug-induced, electrically stimulated, or caused by happenstance—creates images. The fact that the images come in 3-D, as created by the brain or by VR equipment, doesn't make them real. Your personal self-model has been painstakingly constructed out of images from the past stored in your memory. These artifacts of old experiences feel like "you." It's not hard to revisit the times in your life when you made major additions to your self-model. For instance, I can see myself in medical school, on the airplane from India to America, sweating out my first days in a New Jersey hospital as I felt the pressure of the workload, a foreign environment, and the guarded acceptance of American-born doctors. These images pass through the mind as if they were happening all over again—but they aren't.

Self-models are shared at some levels but not others. There's enormous room for personal variations. You and I can spend a day together seeing the same sights, eating the same food, interacting with the same people. A shared self-model would bind us. The Pacific Ocean, a bowl of rice pilaf, and the friends we meet would be part of experiences we share. But your self-model will absorb and reject, interpret and forget, hold on to and let go of the day in an entirely unique way. I may love the ragas improvised by a great Indian sitar player, while you experience the music's microtones as so much garbled noise. If our spouses join us at the table, you and I will be attached to different people with different relationship histories. And so it goes, moment by moment, as the self-model processes every life experience according to its own designs.

# When the Illusion Crumbles

What VR technology reveals is that the self-model isn't limited to one dimension, the visual. We also believe in what we hear, touch, taste, and smell. In effect, you and I were born to mesh perfectly with a wraparound simulation of reality. But there's a hidden ability in the human mind that we cannot overlook. This is the ability to disconnect, to stop identifying with the illusion. Pulling a rabbit out of a hat amazes children because they believe what they see. Once you discover that a hat can have a false bottom, the trick doesn't change, but how you relate to it does. There is no illusion for the magician who has performed the hat-and-rabbit trick hundreds of times. He might feel impatient and bored, wanting to get the whole act over with so that he can eat his dinner. Once an illusion loses its fascination, it loses everything.

The opposite is true, however, when it comes to dismantling the self-model, the wraparound stage set that each of us inhabits. Once you see through this illusion, life suddenly becomes more fascinating. Such is the testimony of people who had the illusion crumble around them, usually without any warning or effort on their part. I recently met someone to whom this happened. Now sixty-eight, Lorin Roche was a broke eighteen-year-old college student at the end of the sixties who agreed to participate in a research project on the physiological effects of meditation. But as recounted on Roche's website, when he arrived at the lab, he was told that "[h]e was a control subject, and received no instructions whatsoever—they paid him to just sit in a totally dark, soundproofed room in the lab for two hours a day for several weeks, and measure his brain waves. With no instructions, and never having heard of meditation, Lorin just attended to the total silence and darkness, and spontaneously entered a state of intense alertness."

This was a startling and unexpected experience, which he found very absorbing while it lasted. A few months later, someone handed Lorin a book of 112 meditations based on ancient Sanskrit sutras (sayings or teachings) from India. He was delighted to find that he had spontaneously had some of the experiences described in these centuries-old texts. For a Western adolescent to have a predilection for meditation is as remarkable as having innate musical talent, but Roche went further. He attended a festival of Bhakti devotees in Joshua Tree National Park in the California desert. In India, Bhakti is the most popular form of worship, consisting of love, faith, and devotion as the path to enlightenment. The most common daily practice is chanting, which Roche was there to participate in.

But the heat was oppressive, and his energy was starting to wane. As he describes the experience, "What sounds good right now is to go jump in a nearby cool salt-water pool. As I get out of hearing range of the festival, I realize the chanting is still going on inside me. And although it is quieter, this internal soundtrack feels powerful. Somehow my atoms are dancing and singing the hymns of praise to the Goddess and the God, Devi and Shiva. . . . It's the Bollywood of the atoms."

Today Roche continues to dance in praise of the Goddess and the God, and he has written a translation of the 112 Shiva sutras that is considered among the most ecstatic. Here are some samples, all based on merging the ancient text with his own personal experiences. Shiva is singing to Devi:

> *Rivers of power flowing everywhere.*
> *Fields of magnetism relating everything.*
> *This is your origin. This is your lineage.*

> *The current of creation is right here,*
> *Coursing through subtle channels,*
> *Animating this very form.*

*Follow the gentle touch of life,*
*Soft as the footprint of an ant,*
*As tiny sensations open to vastness.*

I had an encounter with Lorin Roche while writing this book, and he radiated the blissful state that is the goal of Bhakti—his translation of the sutras is titled *The Radiance Sutras*. I bought a copy to read on an airplane, and its personal authenticity surpasses any other version I know. The "dance of atoms" is real for him:

*Power sings as it flows,*
*Electrifies the organs of sensing,*
*Becomes liquid light,*
*Nourishes your entire being.*
*Celebrate the boundary*
*Where streams join the sea,*
*Where body meets infinity.*

A skeptic would argue that this was such a subjective experience that it has no bearing on reality. The atoms that dance are in Lorin Roche's imagination, not in a physics lab. The physical world doesn't feel like a simulation; it feels totally real, and when we see something fantastic in our imagination, like a flying dragon, we don't run away to escape the fire it breathes.

But this argument misses the point. Everything is mind-made, including fire's heat and its destructive ability. This only shows how complete the illusion is. Fire-breathing dragons are imaginary and a forest fire isn't; as part of the simulation, the physical world operates the way it operates. Fire is hot, ice is cold. Trees burn down, water freezes. The key is identification—once you identify with the simulation, you are embedded in it. You are part of the whole setup, playing a passive role. If your involvement changes, so does your experience. You have a more flexible role to play. Take something as basic as pain. There is no objective way to measure pain. People experience it very differently, and unpredictably.

In a typical pain experiment, participants place their hands in ice water and are asked to rate the pain on a scale of 1 to 10 where 10 is excruciating. Even though the temperature of the water is the same for everyone, one person will rate the pain a 5 (moderate) and another an 8 or 9 (severe to excruciating). On the flip side, if you visit a restaurant kitchen where the pastry chefs boil sugar syrup for candy or frosting, you will observe that many of them can dip their fingers to test whether the syrup has started to thicken, which occurs at well over 212°F. (Women, by the way, seem to have a higher pain threshold than men.)

This result isn't very surprising, but few of us realize that we are actually creating the pain we feel as physically created. As part of the stress response, a rush of adrenaline can block pain, which is why soldiers report that being shot wasn't painful under battlefield conditions, and the same occurs if a person goes into shock. But the total cessation of pain can also occur out of the blue, and with it arrives a dramatic shift in consciousness.

People who have had this experience typically report a "snap" when the mind creates its own altered state spontaneously. In the 2017 book *Stealing Fire*, authors Steven Kotler and Jamie Wheal give the striking example of Mikey Siegel, an MIT-trained engineer who had become burned out with his lucrative job in robotics and artificial intelligence. He wanted more fulfillment, and Siegel began his search by hiking in South American jungles and then visiting ashrams in India, eventually deciding to take up meditation.

On a ten-day meditation retreat, Siegel found himself participating in a focus exercise where the object was to sit still and experience sensations in the body without judging them:

> But Siegel was overwhelmed by sensations. After a week of cross-legged meditation, his back ached, his neck throbbed, and

his thighs were numb. "It was an all-consuming pain," he explains, "and all I was doing was judging."

Moments of extreme experiences, whether pleasant or painful, somehow break the bonds of the conditioned mind, which is trapped in the habit of accepting physical limitations at face value. Suddenly there is access to an ecstatic state, free of conditioning, as was the case for Siegel:

> Something inside shifted. The part of his brain that had been judging suddenly turned off. "It felt like freedom," [Siegel] explains. . . . "It was the most clear, present, and aware I had ever been. And if I could be in extreme pain and still remain peaceful and clear, then I thought maybe other people could do this, too. In that instant, everything I believed about human potential shifted."

Siegel wasn't merely astonished; he didn't let go of this experience, but rather chose to follow up on it. He fervently embarked on the project of "engineering enlightenment," using meditation as one tool among many. For example, we can access calmness by slowing down our heart rate using a wearable biofeedback device. We'll get to the whole topic of interfacing the mind with devices that enhance awareness, but there are some basic points to be made here. Pain has been known to suddenly give way to a detached nonjudgmental state, known as "witnessing." In India, for centuries, sadhus and yogis have undergone *tapas*, or physical austerity, as a path to awakening. The stereotype of the bearded yogi sitting in a remote Himalayan cave reflects one kind of tapas.

Putting stress on the body is found in disciplined Zen Buddhist meditations, where monks get up before dawn, consume green tea and a handful of rice, and then sit in meditation for hours with head and spine erect. As with Siegel, there will predictably be a "snap" moment, when the mind pops out of its identification with pain and the struggle to stop it. But many people are stuck in various ways, and even years of enduring extreme discomfort may not lead to the desired result.

Cessation of pain is powerful evidence that the mind can free itself from sensations that everyone considers a natural aspect of life. But we mustn't miss the wider implications, which carry us to the farthest reaches of reality. Trapped in the self-model, which clings to us more tightly than our skin, we can surrender to it or investigate it. The investigation may be intellectual, the way quantum physics operates. It may use imagination to tease us out of our set ways. When Alice goes down the rabbit hole to Wonderland, the everyday world is reshaped by nonsense, which Alice, being a proper English miss, is impatient with. As Alice watches the Red Queen play croquet using flamingos as mallets or the Cheshire Cat disappear into thin air until nothing remains but his smile, her sensible protests aren't the reader's. We are delighted that Wonderland isn't the everyday world.

Why do we long for wonder? Because we've been there in real life. Wonder existed long before the self-model took over. As one researcher into the mind-altering effects of LSD has concluded, babies don't need psychedelics because they are "tripping all the time." Babies take a while to get with the program, so to speak. They are wide-eyed and delighted with a world that doesn't have to make sense yet. To learn that fire is hot and winter is freezing, a young child must conform to everyday reality. Growing up means learning the rules of the road. But once you learn them, the road turns out to be narrow, and crossing the center line spells disaster. Deviate from the norm and you might just go mad.

Metahuman holds out a third way that is neither as shapeless as the innocence of babies nor as rigid as social conformity. We can live in both realities, in a state William Blake called "organized innocence." Wonder can infuse the everyday world without dissolving it into a trippy puddle. (The famous Indian spiritual teacher J. Krishnamurti, who had a sardonic sense of humor, liked to say that being timeless and eternal, which is part of waking up, doesn't mean that you miss the afternoon train.) The world of the five senses is the organized part. We don't inhabit a chaotic hallucination. The all-enveloping setup that we are entangled in appears to be complete. It covers everything we can see, hear, touch, taste, and smell.

Metareality is the innocent part, where awe and wonder infuse the mind. It's not a mindless state, but it does go beyond rational thought. No less than Albert Einstein confirmed this personally:

> I sometimes ask myself how it came about that I was the one to develop the theory of relativity. The reason, I think, is that a normal adult never stops to think about problems of space and time. These are things which he has thought about as a child. But my intellectual development was retarded, as a result of which I began to wonder about space and time only when I had already grown up.

Einstein never lost his sense of wonder and imbued it with a deeply spiritual quality. "My sense of God is my sense of wonder about the universe," he once said. But it isn't necessary, as I've underscored several times, to couch metareality in spiritual terms. "Going beyond" is an aspect of consciousness, and it is accessible to everyone.

If you ask people how interested they are in investigating reality, not many will respond enthusiastically. But there's a gripping story behind how we got entangled in an illusion. Even more gripping is the possibility of writing a new ending to the story, which has us escaping into the domain of wonder, discovery, ecstasy, and freedom.

# In Your Life

## CHANGING YOUR BODY EXPERIENCE

You are living in an interpreted world, and your body is part of the interpretation. Change the interpretation, and you will experience your body in a new way. When you look at exercise not as a chore, for example, but as a way of increasing your focus and energy, you have created a new interpretation. Now the burn of your muscles on the StairMaster and the windedness you feel after running a mile are positive things, not reasons for distress.

It takes a more basic shift of interpretation to stop seeing your body as a thing, an object suspended in time and space. Is this a mere interpretation? Yes. When you look in a mirror, what do you see? We are conditioned to see a solid, stable physical object with defined boundaries—in that regard, you could be seeing a life-sized mannequin in the mirror. We already know, from discussing the quantum revolution, that matter only appears solid. When you touch your forearm with your other hand—go ahead and do this if you like—it seems as if two solid objects are coming into contact.

In reality, you are experiencing two electromagnetic fields coming into contact with each other, which gives the impression of solidity. For example, two magnets with opposite poles facing each other create a repellent force. If the magnets are powerful enough, a point will come when you cannot push them together until they touch. The repelling force will keep them apart. Therefore, from the magnets' perspective, the air between them feels solid.

The other four senses besides touch also collaborate in the interpreted body. Since photons have no color, the fact that you can see your body as colorful—brown hair, blue eyes, olive skin—is an optical illusion. So are the defined outlines of the body. You do not stop at the barrier of your skin. You travel in a vaguely shaped aura of moisture and exhaled air, trailing behind you a constant stream of microbes and old skin cells that are being shed (by one estimate, the dust bunnies that accumulate in a house are 50 percent dead skin cells). You are also emitting heat and a very mild electrical charge. These emanations have no boundary at all, since they are part of universal fields that extend to infinity.

Nor can you say that you are looking at "my" body, because a question immediately arises: Which body do you mean? Your cells are constantly being exchanged, like bricks flying in and out of a building. The body you see in the mirror isn't the same as it was when you were an infant, or even what it was yesterday and will be tomorrow. Besides the death of old cells and the birth of new ones, atoms and molecules fly in and out by the trillions every hour as your body is nurtured and excretes waste.

The fact is that your body coheres and looks stable, like a building being held together, not by bricks and mortar, but by its blueprint. In your case, the blueprint leaves a physical footprint as DNA, which serves as the template for all life forms. But, once again, the physicality of DNA is an illusion, a mask. The chemical compounds that constitute DNA are phosphates and sugars, and it is only the *arrangement* of them that determines the difference between a banana and the monkey eating it, or between you and a sea snail. These arrangements are nothing but pure information. Therefore, your body is an information construct, and your bloodstream, teeming with thousands of different chemical messages flowing from cell to cell, is an information superhighway.

Having gotten this far, we have dematerialized your body, and yet there's another step to go. What is information? It, too, is a construct. Until the human mind named the construct, information had no formal existence, and some have argued that an information universe could be a kind of quantum soup, swirling, combining, and recombining at every second with lightning speed. This soup can be coded any way you wish.

A physicist could code it in terms of force fields like gravity and electromagnetism. But these fields are unified and merge into the ground state of everything that exists, vanishing from the visible universe into a formless vacuum.

A computer engineer could code the information a different way, as the 0s and 1s of digital programming, but this arrangement of information is only viable for tangible information, like the letters on this page, which you may be reading digitally right now. The mathematics of anything in the visible universe can be calculated. Your DNA is coded by the mathematics of four base pairs (thymine, adenine, guanine, and cytosine) in a sequence with three billion separate units of information. Which brings us full circle, because the base pairs aren't solid matter, either. Even mathematics cannot get at what they are. A mathematical language of 0s and 1s is useful for computer technology, but the immaterial aspects of life—intelligence, creativity, emotions, hopes, fears, and so on—have no mathematical coding. Before Einstein formulated $E = mc^2$, it existed as pure creative potential—a thought not yet thought—and being as yet uncreated, it had no existence in the physical world or the information world or even the mathematical world.

# The Humbling of DNA

The notion that life can be explained by understanding the human genome has prevailed for decades, but in fact your DNA turns out to be a bit player in the larger scheme. The failure of DNA to explain how life emerged is startling, although the general public hasn't heard about this very much. This is a perfect example of how materialistic explanations always fall short, so it's worth telling the story in some detail.

The accepted story, which everyone learned at school, is that DNA contains the "code of life," a master blueprint that jumps into action the instant an egg is fertilized in the mother's womb. From that point on, a human being develops from a single cell to 30 trillion cells as the blue-print unfolds. As powerful as the "code of life" story is, behind the scenes a growing number of geneticists don't buy into it; in fact, they think we've gotten a lot about genes wrong. In various ways the "code of life" has huge holes in it that are growing bigger every day. This is outlined in an online article in the journal *Nautilus* titled "It's the End of the Gene as We Know It." The author, Ken Richardson, is an expert in human development, and he gives us a remarkable view of how cells work, which depends much more on invisible ingredients like intelligence and creativ-ity than on molecules, even one as complex as human DNA.

Richardson's argument goes as follows: DNA's purpose is to produce the proteins that are the basic building blocks of a cell. But DNA alone does not account for the many ways that cells, tissues, and organs use these proteins. The notion that DNA contains the blueprint for the body is basically dead in the water. Recent research has shown that cells are dynamic systems that change their makeup "on the hoof," as Richard-son puts it, a process of self-regulation that begins almost the moment a sperm fertilizes an ovum.

As soon as that one cell forms into a minuscule ball of identical cells, Richardson writes, "[they] are already talking to each other with storms of chemical signals. Through the statistical patterns within the storms, instructions are, again, created de novo [i.e., from scratch]." It turns out that totally independent of DNA, a cell is controlling all kinds of in-formation contained in amino acids, fats, vitamins, minerals, enzymes, various kinds of nucleic acids (RNA)—a whole factory of ingredients necessary to keep the cell going is not predetermined by our genes at all. This self-regulation implies tremendous intelligence.

In the newly emerging view, the cell controls DNA just as much as DNA controls the cell. The situation has been like this from the begin-ning of life on Earth. DNA, it seems, emerged at a late stage of cellular evolution. In their earliest stages, billions of years ago, cells had no DNA but were self-enclosed vats of molecular soup. This soup somehow began to regulate itself, giving rise gradually over time to permanent structures that were needed on a regular basis, such as proteins, enzymes, and probably RNA, which makes proteins. The information for these structures was then coded as DNA, which serves as a kind of passive database. Richardson notes something else that puts DNA in its rightful place: "More startling has been the realization that less than 5 percent of the genome is used to make proteins at all. Most produce a vast range of different factors (RNAs) regulating, through the network, how the other genes are used."

As validation of this new understanding, it is now known that cells can alter their own DNA—this has emerged in the new field of epigenetics, which explores how everyday experience leaves chemical "markers" on a gene, altering how it functions. Far from robotically following a fixed blueprint, the life of a cell is highly dynamic and flexible, responding to changing conditions on a microscopic scale. If this wasn't so, we couldn't respond to life on a macroscopic scale.

Being human means that we think and act creatively, using our intelligence to devise new ways of meeting all kinds of challenges. DNA didn't discover fire or invent the personal computer. The fact that DNA is responsible for the manufacture of proteins is important, but it's seriously mistaken to expand its role to life as a whole. Richardson is particularly worried that wildly exaggerated assumptions about DNA could lead to social policy that echoes the racism that fueled the eugenic movement decades ago, most notoriously with the Nazi ideology of a master race. As a case in point, Nobel laureate James Watson, who co-discovered the structure of DNA in 1953, was recently stripped of all his honors at Cold Spring Harbor Laboratory, where he spent much of his scientific career, after he continually expressed his bigoted opinion that black people and women are less intelligent than others based on their genetics.

With the blueprint of life crumbling before our eyes, what next? At present the new story in genetics is stuck on two factors, information and complexity. The notion is that primal "molecular soup" found ways for atoms and molecules to form complicated structures through information exchange and the statistical possibilities that arise when zillions of molecules start churning around. But is that feasible? Can the human brain, for example, be the end product of swirling soup to which more and more "stuff" is added? As someone wittily put it, the notion that complexity is enough to explain the brain is like saying that if you add enough cards to the deck, they will start playing poker.

Because science is tied into a materialist explanation for everything, it has an enormous blind spot. A cell biologist cannot make the leap to invisible traits every cell displays beyond its chemical structures, namely, intelligence and creativity. Logical analysis has been science's most powerful tool, and it's no small accomplishment to replace myth, superstition, and popular opinion with rational facts. Is it really possible, though, that a sudden creative leap comes about because someone followed the rules of logic? The obvious answer is no, and, as proof, we can offer the amazing imaginative leaps made by the quantum pioneers who uncovered the absolutely illogical quantum domain. More recently, the existence of dark matter and energy revealed another domain, even more peculiar and illogical than the quantum world, which doesn't even interact with ordinary matter and energy. Your body isn't a machine governed by logic, which is why any attempt to turn it into a kind of super-complicated machine is bound to fail. Too much of the information being sent throughout the body, affecting all thirty trillion cells, is generated by emotions, hopes, fears, beliefs, mistakes, and imagination—all the most important things that give richness to human existence.

Matter and energy behave very peculiarly at the quantum level, to the point that solid physical objects are undermined. Every phenomenon in the universe can be reduced to ripples in the quantum field as it interacts with the gravity field or the more arcane quark field. On the surface of life, solid objects are just slow-moving ripples compared with, say, photons traveling at the speed of light. Physics can back and fill by pointing out that the human body, like all solid objects, remains intact despite all the quantum funny business. But your body is intact due to yet another field, the electron field.

This reassurance only holds true, however, as long as the information in DNA is intact—with physical death, electromagnetism hasn't changed, nor have the atoms and molecules that constitute your body. But the process of decay breaks down the invisible bonds of life. Cells lose the real glue that makes life possible, which isn't electromagnetism. No one can say with logical certainty why the human body doesn't fly apart into a cloud of atoms that the next breeze will blow away.

It's easy to have your head start to spin when you realize that your body, at best, is a constant stream of ever-shifting information, but we mustn't lean on this as a crutch for keeping the physical world intact. Information, remember, is a human concept, like any other model. To say that we are intact because of information has its limits. It's not as if 0s and 1s are sticky. They don't glom on to one another. The way that 0s and 1s get glued together is through human interpretation. We know that information exists because we invented the concept.

So where did we get this ability to glue the world together and give it meaning? The answer will be persuasive only if we can apply it to our bodies. How did we acquire the ability to hold our bodies together? That ability must lie outside the body, because we can't say that our bodies told us how to live and be and think. We can't even claim that our brains told us how to live and be and think. The brain is another physical object, and it would be circular logic to say that a physical object created itself. (In the field of artificial intelligence, this is like saying that there was a robot that invented robots.)

No matter what angle you take, the body vanishes into the realm of concept, mind, and intangible agencies that are its true creator. But concepts, mind, and immaterial agencies must have a source. Before you can paint the *Mona Lisa*, there has to be the concept of art. What gave rise to art that isn't art already? What gave rise to concepts that isn't a concept already? The only possible answer, as this book argues from many per-spectives, is consciousness. There is no other building block that viably explains all the mysteries we've just touched on, from creativity to hold cells together, to how inanimate atoms and molecules somehow arranged themselves into living creatures.

There is much more to say, yet if you look at your reflection in the mir-ror, you can already see that it is only a solid, stable "thing" with defined edges because you interpret it that way. My aim isn't to plunge you into a state of confusion over your body, however. My aim is to free you from all interpretations that force limitation upon you. Being human can only be defined as limitless. When we impose limitations, we diminish being human. That's the truth of metareality, a truth we can inch toward step by step until it becomes a living reality for as many people as possible, including you and me.

# 2

# "I" IS THE CREATOR
# OF ILLUSION

When you see your reflection in the mirror, the fact that you recognize yourself comes naturally and seems too basic even to comment on—yet this small act of self-awareness turns out to have tremendous significance. The self you have learned to recognize in the mirror is constantly reinforcing all kinds of limitations that do not need to exist. When William Blake spoke of "mind-forged manacles," he could just as well have said "ego-forged" instead.

It's impossible to remember a time when you didn't look in the mirror and see yourself. But there are steps in childhood development that gave you your first inkling of "I," your sense of self. Mirrors hold no interest for very young infants, for example, and, surprisingly enough, walking and talking precede the time, around eighteen months, when a young child recognizes that he is seeing his own body in a mirror. After that, it becomes a favorite toy. (The few animals that can see themselves in a mirror also become fascinated with their own image once they catch on.)

At the very least, we need to realize that no one lives in the same reality. Everyone's version is personal. A hundred people viewing a glorious sunset in Hawaii are actually seeing a hundred different sunsets. For a person who's feeling depressed, there might be no beauty, much less glory, in any sunset. Since "I" is central to every person's version of reality, it's a key element in the simulation we accept as real, and until we can know ourselves beyond "I," the illusions of virtual reality will keep us in its grip.

*Illusion* is a loaded word. Society disapproves of someone who is under the illusion that nobody else in the world really matters; we call this an inflated ego or solipsism. But the illusion that love conquers all, which everyone believes if they happen to become deeply infatuated, is an illusion we'd all like to believe in all the time—falling out of love, which replaces illusion with reality, is quite painful. A mix of pleasure and pain characterizes "I." On the pleasurable side, discovering their identity makes little children exultantly happy. The "terrible twos" reflect a rampant display of egotism, where the child asserts, "This is me! Pay attention. I am here!"

The terrible twos are notorious for being a maddening time for parents, because the naked assertion of ego is obnoxious. More important, it's unrealistic. You can't survive in society if you run around demanding that the world pay attention to you all the time, or even most of the time. Adult life is a compromise between getting what you want and going along with social norms, between an all-consuming "I" as the center of the universe and a muted "I" that's a small cog in the vast machinery of society. The balance isn't easy to live with, and countless people fall into the trap of feeling insignificant, while a few are allowed to aggressively impose themselves on the rest.

Psychologists spend their careers mending people's damaged sense of self, but on the road to metahuman, we must ask a more radical question: Why should "I" exist in the first place? It delivers a life of unpredictable pleasure and pain. It isolates us from the world and limits what we feel, think, say, and do. How often are we held back from doing something impulsive because we automatically think, "I'm not the kind of person who does X"? That X can be anything from pulling a practical joke to bragging about how much money you earn to running off and joining the circus. Every limitation imposed by "I" is actually pointless. It serves only to uphold old conditioning from the past.

When we see that "I" is a mental construct—and a very shaky one at that—it becomes open to change. We might decide to do without it entirely once "I" no longer serves its purpose. "I" exists to convince you that you are a creature of virtual reality, and that going beyond the simulation isn't possible, any more than a portrait can jump out of its frame. The reason we find ourselves hopelessly entangled in an illusion is that we are totally wrapped up in "I" and everything it stands for.

There's a lot of dismantling to do before "I" stops ruling a person's life. From our first memories of having a self, "I" has been our closest companion, and it spends every waking moment glomming on to desirable experiences and kicking away undesirable ones. "I" doesn't want to give up its power, and for good reason. Having one special person love you and only you makes life worthwhile. When "I" fades away, who is there to love and be loved? But there's much more at stake. Everything a person thinks, feels, says, and does is in service to making "I" stronger, happier, and better. Becoming metahuman cannot succeed unless it offers something more fulfilling than anything "I" holds out to us.

# The Ego's Agenda

At first glance, the ego looks indispensable. How can we abandon something we need to survive? "I" is the reason you feel like you and no one else. You gaze at the world through a pair of eyes no one else possesses. A mother spotting her child as he comes out of school to be taken home receives the same visual information as every other parent waiting in the parking lot, but she literally sees a unique child, her own. Uniqueness is precious, but it comes with a price. Almost none of us is comfortable being totally on our own, and the prospect of becoming an outcast is very real if you insist on being yourself. The poet William Wordsworth rhapsodizes, "I wandered lonely as a cloud / That floats on high o'er vales and hills," but very few of us view *lonely* as a positive. When somebody is so selfless that they surrender all personal needs, they're sometimes called saintly. It's more likely, though, they they'll be labeled antisocial or mad—it's hard to believe that someone can be normal and yet totally free of the ego and its need for pleasure and approval. Many spiritual movements denigrate the ego as a burden, a curse, or a hidden enemy of higher consciousness.

Ironically, calling the ego your enemy is an ego judgment. Calling the ego your friend is also an ego judgment. Therefore, to say, "I want to be without ego" is self-contradictory; the ego is saying that, and it certainly doesn't want to commit suicide. Your very words cannot get you to a place outside the illusion you are entangled in. You can't pluck out the ego like removing an inflamed appendix. If you think you can, you only sink deeper into the illusion by fooling yourself that you are selfless. "I" is a tiny thing, a single letter. But what you have built around it—what everyone has built around it—is like a coral reef made up of minuscule cells hardened over with a massive shell.

If this description sounds extreme, consider how you process raw experience. Experiences are interpreted and become part of how you personally accept or reject reality. We don't witness how this occurs because most experiences seem too insignificant to matter. For example, you might taste a vindaloo curry in an Indian restaurant, find it blazing hot, decide you don't like it, and never order it from a restaurant menu again. Someone else, raised in the Indian state of Goa, where vindaloo is a staple dish, barely registers the heat of the chilies that go into the recipes and instead has nostalgic memories about his mother's vindaloo.

The two experiences, as raw data entering the brain through the sense of taste, appear to be identical. But they aren't—experience always passes through someone's personal interpretation. "I" is having every experience, not the five senses or the brain. Reducing experience to raw data is totally misleading, as if the eardrum determined which music you like, or brain cells decided that a Rembrandt painting was a masterpiece. "I" makes all such decisions, and, as it does, every experience makes the power of "I" stronger.

Experiences, by nature, are fleeting and momentary. As soon as I finish saying "thank you" or take a bite of chocolate or kiss my grandchild, the experience has vanished. Based on this undeniable fact, you have two options. You can accept how fleeting every experience is, or you can hold on to it. When you choose the first option, life is a flow of fresh experiences, like a stream constantly renewed at the source. You are not haunted by bad memories or filled with anxiety about what might happen next. If you pick the second option, you accumulate a storehouse of habits, conditioning, likes and dislikes, and a catalog of things you never want to repeat again. The second option is the foundation of the ego, which holds on tightly to reinforce "I" and its sense of security. The loss is very great, however, because experiences won't stop occurring just because you want them to, and by holding on, you shut out the flow of life.

What drives us to hold on instead of letting go? A simple fact of life: "I" has an agenda. The illusions created by the mind are not random. "I" is in charge of your self-interest, and its agenda serves one demand— "More for me." We shouldn't be surprised when "More for me" becomes insatiable, as billionaires crave more money and despots more power. The average person cannot relate to such extremism. But the need for more is powerful in everyone, because everyone has needs and desires that want to be fulfilled. We all need security and a sense of safety. To need love makes us human. Needing to explore the world is an unstoppable urge in a toddler careening around the house and getting into everything.

But look deeper, and it becomes clear that "I" is *based* on need. It ties you to a program of constantly finding new needs that never end, which is the opposite of fulfillment. Fulfillment is the state of needing nothing because you are enough in yourself. Consumer society promotes neediness as normal—there is always something new to buy that will at last put a satisfied smile on your face. Thus, a normal life is actually a life of lack desperately and constantly trying to fill a black hole that will never be filled. When you are needy, fulfillment is unattainable.

Here a very important insight dawns. "I" doesn't have an agenda. "I" *is* the agenda. The ego comes with built-in demands, no matter how hard we struggle either to deny these demands or to fulfill them. Neediness is a state of awareness, and "I" will never loosen its grip until we find a higher state of awareness.

# A Mysterious Birth

"I" creates obstacles that keep metareality shut out, as if by a thick wall, even though the wall is invisible. It's important, in our journey of self-awareness, to understand why human beings chose to isolate themselves in this particular way. Was there a time when "I" was weak or didn't exist? Even though ego is now an ingrained part of the human psyche, it has a history. It has left clues in physical form, like a trail of footprints in the forest made by an invisible creature. For example, one sign that you are an individual "I" is that you answer to your name. The first name is lost in prehistory, but the first written name belongs to an Egyptian pharaoh, Iry-Hor (the Mouth of Horus) from 3200 BCE.

Once you begin to investigate, other clues about the evolution of self-awareness emerge. Much earlier than written names came the ability to recognize our reflection. We have no way to re-create what our remote ancestors experienced, naturally. Did prehistoric humans gaze into dark pools of water and recognize their reflection? The speculation is that they did, but the event cannot be dated. But the invention of mirrors came along very recently, measured in evolutionary time. Polished stones used as mirrors date to 6000 BCE in what is now modern-day Turkey, and as ancient civilizations emerged in Egypt, South America, and China, whatever could hold a polished surface, from obsidian and copper to bronze and silver, was employed for this purpose.

Are we the only creatures who can see themselves in a mirror? A pet parakeet will play with its image in a mirror because (we assume) it sees another parakeet there. Dogs and cats typically show no interest in mirrors. But, oddly enough, self-recognition evolved in creatures that have no reason to possess this ability. Chimpanzees, gorillas, and other great apes do see themselves in a mirror. How do we know if a creature actually sees itself in a reflection? The most telling test is actually quite simple: put a pink hat on the animal's head. When the animal looks in the mirror and sees the pink hat, does it touch the hat on its head or the hat in the mirror? If it touches the hat on its head, it passes the test of "That's me I see."

Yet great apes don't have mirrors in their native habitat, so there seems to be no evolutionary reason for this ability. Likewise, we don't know why three other creatures—magpies, elephants, and dolphins—can recognize themselves in a mirror. Magpies use their reflections to preen themselves more, while elephants, once they understand how a mirror works, engage in novel new behaviors. For example, they spend an inordinate amount of time examining the inside of their mouths, an area of the body they couldn't see without the aid of a mirror. (If this topic fascinates you, go to a YouTube video that shows Asian elephants and how they behave when confronted with a mirror: https://www.youtube.com/watch?v=-EjukzL-bJc.)

Mirrors aren't the only way we recognize ourselves. The oldest artifacts that hint at self-awareness are sculptures that depict humanoid forms. What makes them so astonishing, according to the most recent archeology, is that such objects predate the rise of *Homo sapiens*. A simplified time line will help us to get our bearings:

| 14 million years ago | First great apes appear |
|---|---|
| 2.5 million years ago | The genus *Homo* evolves |
| 1.9 million years ago | Hominids evolve into *Homo erectus* |
| 200,000 years ago | *Homo sapiens* appears |
| 10,000 years ago | End of the last Ice Age |

By the time *Homo sapiens* was becoming a distinct species, around 200,000 years ago, our closest ancestor, *Homo erectus*, had long discovered fire and toolmaking. Nor did hominids wait for our species in order to develop a sense of self-awareness. In extremely ancient ruins have been found crude humanoid figures fashioned by *Homo erectus*. They are astonishingly old. The first to be discovered was the Venus of Berekhat Ram, a basalt artifact unearthed in 1981 by a team of archeologists from Hebrew University on a dig in the Golan Heights near Syria in Israel.

The Venus of Berekhat Ram consists of two round shapes, the bigger one suggesting a body, the smaller one a head. Three incisions can be seen, two on either side of the "body" standing for arms, one encircling the "head" without standing for any facial feature. Despite the name *Venus*, this object is so primitive that at first some experts believed it was an accidental formation made by natural erosion. The debate over whether these are intentional marks etched by an artist was settled when a closely related figure, the Venus of Tan-Tan, was later discovered in Morocco. The two sculptures resemble each other so closely that they could have been created by the same hand.

Dating the Israeli find was exciting but difficult to pinpoint. The Venus of Berekhat Ram was sandwiched between two layers of volcanic deposits, one from around 230,000 years BCE, the other from 700,000 years BCE. The sculpture was made sometime in that vast expanse of time. To the modern eye, the Venus of Tan-Tan from the same Lower

Stone Age looks more convincingly human, since it has a torso, head, and legs. The fact that a mind predating not just *Homo sapiens* but Neanderthals felt the urge to depict its form artistically is a sign of self-awareness woven into the very fabric of our existence. The sculptor is saying, "This is what I and my kind look like." As far back as anyone can tell, there were never humans *without* self-awareness.

"I" hasn't just survived since prehistory: it metastasized. All around us we see evidence of malignant selfishness. The grotesque excesses of greed in our present Gilded Age is a symptom of "I" run amok, and we've seen how recklessness in the financial sector can bring about disasters in the global economy without the moneyed culprits bothering to care—or stopping in their pursuit of even more wealth. If it weren't for the ego's drive to defeat other egos, to make itself important by denigrating anyone who is different, there would be no need for us-versus-them thinking and the endless conflicts this has created, from family squabbles to civil wars, religious crusades, and the global atomic threat. Can we account for this metastasis and come up with a cure?

If you lived through the Cold War and the threat of nuclear devastation, you have seen how "I," having formed an enemy, will carry enmity to the brink of mass destruction. Even if the nuclear shadow somehow vanished, nations would continue to perfect new, more deadly means of mechanized death. It would benefit humankind to reduce the amount of damage we do to ourselves that is directly traceable to our habit of viewing the world from the ego's perspective, given the needless fear and suffering it has brought.

# Choosing to Be Separate

No one catches pneumonia or even the common cold voluntarily, but when it comes to "I"—whose ill effects reach into every corner of life, we have chosen to be separate—this is a species trait. We have evolved to feel superior to every other life form. On the one hand, this gave us a major evolutionary advantage. Consider how we relate to the environment. Every other creature adapts to the environment and merges with it. Over billions of years, evolution has created exquisite mechanisms for adapting to the most inhospitable reaches on the planet. The interior of Antarctica, for example, contains a specific kind of mountain known as a "nunatak," a peak that crops up from the thick surrounding ice cap. A more desolate environment would be hard to imagine, with nothing but ice fields in all directions, subzero cold, howling winds, and seemingly no foothold for plants or animals.

Yet there are records of a white seabird known as the snow petrel (*Pagodroma nivea*) nesting in nunataks as much as 60 miles inland from the coast, to which they must return to skim the water for food. When mating season arrives, snow petrels find exposed rock crevices for their nests made of small pebbles, and a mating couple nurtures a single egg in a frozen wilderness for forty to fifty days before it hatches. Evolution placed the snow petrel in this situation, but humans have a choice about where and how to live.

These choices weren't dictated by our physical limitations. Humans have encroached on the planet's farthest reaches much more than our hominid ancestors could physically endure. It is our force of will, an inner drive that is determined to bring Nature under control, that impelled us to inhabit all but the most lifeless environments in terms of extreme heat and cold, scarce food supplies, long periods of the year without the sun, high altitude, and so on.

When we were still in our naked state, extreme physical hardship did push us to the edge of survival, nearly extinguishing *Homo sapiens* almost as soon as our species appeared. It took awareness to overcome the physical odds against us. As detailed in a 2016 *Scientific American* article, titled "When the Sea Saved Humanity," human survival was touch and go, and most of our ancestors didn't make it. The article's author, Curtis W.

Marean, is an archeologist from Arizona State University, whose team discovered the evidence for this evolutionary crisis. Marean writes:

> At some point between 195,000 and 123,000 years ago, the population size of *Homo sapiens* plummeted, thanks to cold, dry climate conditions that left much of our ancestors' African homeland uninhabitable. Everyone alive today is descended from a group of people from a single region who survived this catastrophe. The southern coast of Africa would have been one of the few spots where humans could survive during this climate crisis because it harbors an abundance of shellfish and edible plants.

At caves along a section of the South African coast, known as Pinnacle Point, archeologists have found abundant mollusk shells and occasionally remains of seals and whales, indicating that almost fifty thousand years before previously explored sites, early humans had learned to harvest the sea for food while the harshness of an Ice Age climate caused almost everyone else to succumb. Tools in the caves suggest that these survivors had high cognitive abilities—Marean makes a strong if controversial case for mental faculties that were totally necessary for survival, such as calculating tidal rise and fall by the moon. Only at low tide, he says, could the inland cave dwellers trek to the sea and undertake the hazardous venture of harvesting mussels and other mollusks against the pounding surf.

Trapped in the direst distress, our ancestors had no avenue of physical rescue or escape. How did they find the means to rescue themselves?

## Editing Reality

The answer is not physical. Fascinating as these archeological discoveries are, it wasn't harsh external pressure that forced our ancestors to adapt. It took a great reality shift "in here." We became a consciousness-based species, using the mind to outwit Nature's challenges. One of the most important factors in the expansion of human consciousness was that our brains became too big, efficient, and complex for their own good. A kind of brain overload fueled our desperate need to whittle it down so that daily life would be manageable. If the rush and hubbub of a modern city seem like overload, that's nothing compared to the mental crisis our remote ancestors faced.

The problem wasn't that the human brain simply grew and couldn't stop. The problem was that instinct, which guides how other creatures behave, began to dwindle in us. A honeybee seeks only flowers; it instinctually stings an intruder; only the queen bee lays eggs. Human beings have a choice in all three behaviors. We explore Nature for all manner of food. We fight or keep the peace in different circumstances. We mate according to extremely complex behavior patterns. Having been freed from instinct, the choices that face us are literally infinite. The brain cannot be infinitely large, however. So how can the human mind fit infinite choice within a finite physiology?

This wasn't a dilemma faced only by our remote ancestors. Every new-born baby comes into a world where too much information is constantly bombarding the higher brain, a flood of raw data that could never be processed in its totality. Think of searching for your car in a crowded parking lot. To find it, you don't visually take in the pavement, sky, people, and every vehicle, still or moving, in your field of vision. Instead, you have a mental image of your car, and, with focused attention, you edit out everything that's irrelevant to one task, finding a specific vehicle.

This points to another reason we develop an ego. People identify with what they can do. A car mechanic is different from a concert violinist. A sentence that begins with "I am X" can end with all kinds of behaviors, traits, talents, and preferences. By the same token, a sentence that begins with "I am not Y" can also end in many ways. As it turns out, the list of things that we choose not to be is much longer than that of the things we choose to be. If you are a Christian, that's a single choice that excludes all other religions—at present count, there are 4,200 faiths in the world that a person with one faith doesn't have to think about except in passing. As we exclude countless choices without even thinking about them, we are editing reality according to the dictates of the individual "I."

This ability to edit raw reality was already present in animals that hunt for a specific prey, for example, but there was no conscious choosing involved. When penguins and other seabirds that nest in huge colonies come back to shore with their crops full of food, they somehow find the one specific fledgling that belongs to them in the overwhelming din of noise created by thousands of chicks. The arctic fox can detect the movement of mice underneath several feet of snow in the winter and pounce precisely on its prey. Monarch butterflies can follow an exact migration pattern to and from one locale in Mexico where they breed.

There's an enormous mystery about how humans developed focused attention, not for specific types of food and locales, but as a trait we can turn on and off. The things you're interested in fascinate you and hold your attention, while the things you have no interest in escape your notice. The appeal of detective novels lies in how cleverly a Sherlock Holmes notices the tiniest, seemingly irrelevant clue. (Holmes, we are told, was an expert in cigar ash and what kind of tobacco each ash represented, but he didn't know that the Earth revolves around the Sun because that bit of knowledge was useless in the art of crime detection.)

Even though we cannot solve this mystery of paying attention versus tuning out, there's no doubt that "I" is in charge of both. My wife and children and grandchildren are objects of deep personal interest to me (the kind of interest we label as "love") while they are total strangers to nearly seven billion other people on the planet. Once attention is focused, emotion follows. Growing up, my son Gotham loved the Boston Celtics basketball team and hated the Los Angeles Lakers. This became part of who he was, an either/or decision that he identified with.

Either/or is the most basic editing tool our minds possess, and it begins with "me or not me." Ego separates each person from every other person though countless decisions about "me or not me." Many such decisions have no real purpose except to reinforce the ego. (It's not as if Celtics fans are better, smarter, or more well-off than Lakers fans. Yet when Gotham found himself moving to Los Angeles, and his work put him in close contact with the Lakers, it was a wrenching change. Turning "not me" into "me" can be very hard. Imagine, for example, that you had to spend a year working for a political party you've disliked all your life.)

As the ego metastasized over time, the "other" acquired differences became the basis of social suspicion and disapproval. Before my children were born, I was a newly arrived immigrant working at a hospital in New Jersey during the doctor shortage caused by the Vietnam War in the 1970s. Every day going to work, I knew in the back of my mind that the American-born doctors in the emergency room looked upon me as their inferior because I had come from India.

If we stand back and consider the whole picture, "I" edits reality too much and for selfish reasons. We deliberately close ourselves off from new possibilities to suit old, fixed preferences. Everyone's past is a chaotic collection of choices about what they like and dislike, how they feel emotionally, and the memories they carry around as baggage, not to mention their fixed beliefs, family history, and every life-altering experience since birth.

You weren't shaped by what happened to you since birth. You were shaped by what you thought about those happenings. The ego and every response it has ever had is a vast mental construct—the metastasis of "I"—that grew from the seeds of ego in our remote ancestors. Our ability to edit reality is responsible for everything a human being can decide to pay attention to, and since we pay attention to billions of things, reality in its unedited state must be vastly larger. Human achievements represent a tiny fraction of what reality has to offer—the horizon spreading before us is unlimited.

# Deciding to Let Go

We've been covering what might be called the natural history of "I," and it has told us many important things about how the virtual reality we accept as real is actually mind-made. In the human mind, reality is constructed so that

The available information isn't overwhelming and chaotic.
We feel free to accept or reject any aspect of reality we choose.
We seek to repeat the most familiar, safe, and agreeable
    experiences.
We avoid the most threatening, strange, and disagreeable
    experiences.
The ultimate judge of what's real is the ego, which is highly
    personal and selective about how we interpret the world.

I'm not here to declare that the ego is your enemy, which would be just another ego judgment. From a neutral standpoint, the ego is limiting. Having chosen to travel through life with "I" as your most intimate companion, you have silently agreed to filter, censor, and judge your experiences. That's the primary use of consciousness in most people's lives, and it's like using a powerful computer just for emails. To limit your personal reality shuts you off from the infinite potential that is the greatest gift of consciousness.

At some point "I" edits out too much reality or misses the important things that could expand love, compassion, creativity, and evolution. We spend too much mental energy focusing on things that are damaging and self-defeating. If you've attended a Thanksgiving gathering where the same tired, vexing family issues are hauled out year after year, you know how stubbornly "I" can cling to petty, irksome things. For creatures trapped by physical evolution, there's no escape. Cheetahs are the fastest runners on Earth, but their amazing speed has made them smaller and weaker than other predators. The most vulnerable stage in their life cycle is at birth, when the cheetah mother is limited in her ability to protect her young. It is estimated that 90 percent of newborn cheetahs do not survive. Added to this is the speed of the gazelles that the cheetah chases as its prey. The gazelle and the cheetah are so closely matched that adult cheetahs often fail in pursuit of food and therefore live on the brink of starvation. Trapped by their specific evolutionary adaptation, cheetahs can't turn to the other foods that are all around them—termites or grass or mice—to fend off starvation.

*Homo sapiens* faces the opposite predicament. Our minds open up the field of infinite possibilities. Having the ability to bend Nature to our will, we make deliberate choices that seem beneficial to our survival, but decisions have unforeseen consequences. Defending themselves with weapons advanced the rise of early humans, and a weapon as sophisticated as the bow and arrow appeared as early as 45,000 BCE. Then weaponry couldn't be stopped, making the catastrophe of the nuclear arms race inevitable. Or was it? Freedom of thought is our natural state; being trapped by the past isn't. We are still in that state of liberation, should we choose to take advantage of it. The pivotal issue is the metastasis of "I," which has taken free will too far in the service of anger, fear, greed, blind selfishness, and all the rest.

Once we see this, we can understand how personal relationships get sabotaged. Two people fall in love and get married. After the honeymoon, they must relate to each other in all kinds of ways—doing household chores, making money, scheduling time for things to do together or apart—and "I" does its job of managing one situation after another. But if you are having a fight over family finances, your ego brings up anger, the need to win, and the stubborn desire to be right. If the argument gets heated enough, grievances over older wrangles come bubbling up to the surface. Unless you are careful, a trivial disagreement gets bitterly personal. What has been lost in the heat of the moment is the underlying love that sparked the relationship in the first place. That's the larger reality, which "I" single-mindedly excludes so that it can win a small and usually pointless argument.

Two people occupy a small dot on the map. Now expand the territory on a global scale. The human race is ravaging the planet because seven billion people, acting on the advice of "I," prefer local experience over solving a global problem. Wars break out and populations incur death and destruction on a massive scale because the larger territory—maintaining amicable peace—is sabotaged by the anger generated by every "I" choosing to follow its irrational, angry, hostile agenda.

The bottom line is that "I" firmly believes it can manage reality, and yet human history is littered with its abject failures. Even the basic assumption that "I" is in contact with reality is false. At this moment you have no actual experience of the quantum field, from which everything in creation springs. You have no experience of the atoms and molecules that constitute your body, nor of the operation of your cells, nor of the brain itself. It seems strange that the human brain has no idea of its very existence. Viewing a brain under the surgeon's knife or while dissecting a body in medical school is merely the secondhand observation of a mushy gray thing with grooves running across its outer surface. Nothing observable hints that this mushy stuff processes consciousness.

At bottom, "I" polices our experience to make sure that life remains local and not infinite. Infinity is the ego's enemy, because infinity is the whole map, not just dots and pins stuck in it here and there. To let go of "I" is to embrace infinity. Only by being comfortable with our infinite potential can we discover that reality doesn't need editing. Wholeness is where we belong. Once we begin to chop wholeness into bits and pieces, the ego takes over to manage each one, bit by bit, and, whether we realize it or not, it depletes us physically and mentally. So we need to investigate whether infinity is a livable environment. If it is, then letting go of the ego can be justified. And no matter what "I" has done to improve life, we may begin to realize that living in wholeness is better.

3

———

# HUMAN POTENTIAL
# IS INFINITE

Human potential is infinite because consciousness has no boundaries. Being human means that anything can happen. For the moment let's explore the inner world of possibilities. There are ways in which outer reality is much more malleable through consciousness than anyone supposes. Since consciousness is the foundation of reality, we shouldn't set down any absolute limits. In order to fly, human beings didn't sprout wings, but we did find a path to achieve what we envisioned. Keep in mind that a path will always be open, no matter how far-fetched our aspirations.

In the inner domain of consciousness, the possibilities for new thoughts, insights, and discoveries is already unlimited. This alone makes it important to see *Homo sapiens* as a consciousness-based species. Infinite possibilities are part of our makeup. But something inside us resists believing in infinity as a human quality. Edited reality feels more comfortable. There are extraordinary events, however, that prove, quite literally, the reality of the notion "Anything can happen." Sudden leaps in consciousness are occurring all around us if we only take the time to look. But "Anything can happen" could describe a totally random universe filled with quantum noise. In quantum theory, everything bubbles up in "quantum foam" before form and structure begin to appear, rather like formless cake batter until the cake is baked and acquires a solid shape. Because quantum particles wink in and out of existence, according to the laws of probability, there is an infinitesimal chance that a physical object—a leaf, a chair, or a sperm whale—could suddenly appear out of the blue. Physics exists to tell us exactly how nearly impossible such an event actually is.

You might think that these matters are so abstract that they have nothing to do with you personally. When quantum physics asserts that the chance of your driveway being blocked by a giant squid is very, very tiny, you aren't learning anything new. We participate in the world "out there" the way society taught us to. But this is a radically short-sighted view of reality. Our participation in the world actually begins at the quantum level. For all practical purposes, this is where mind and matter meet. Both exist as possibilities, ready to emerge into manifestation but still invisible.

Therefore, mind and matter are more malleable there, just as soft clay is more malleable than a china plate or figurine after it has been shaped, fired, and glazed. Being human, we can consciously return to the quantum level, which expands our participation infinitely compared with what we can do once creation has hardened into place. Not only is creation softer, so to speak, but mind and matter haven't separated yet. When they do, a rock won't instantly become a rock while a thought goes another way and instantly becomes a thought. Mind and matter first emerge exactly alike, as invisible ripples in the quantum field. These ripples will form patterns as they run into ripples from the gravity field, the quark field, the electron field, and a few others. Interference patterns, which are like the rippled imprint of waves left on a sandy beach, get set up, and only then do recognizable objects such as quarks and electrons begin to appear.

Physics has done a brilliant job tracing creation back this far. It has done very little, however, to trace the mind back to its source—we'll have a lot more to say about that. But the upshot is that the average person assumes that mind and matter are naturally separate. That's how the illusion confuses us. Thinking about an apple is very different from holding a physical apple. Yet at a deeper level the thought and the apple were once the same. They begin as possibilities in the domain that physics calls virtual reality. The mystery revolves around how two things that are so unalike could have come from the same seed.

To take advantage of infinite potential, you must accept that reality is open-ended, capable of taking subtle, invisible impulses and turning them into mind and matter. This open-endedness I'm calling metareality, because it lies beyond mind and matter; metareality is where the entire universe and everything in it, including all mental activity, is in an embryonic state.

I won't pretend that metareality is easy to embrace when you first learn about its existence. How secure can you feel about your life if it is totally open-ended? Not very—everyone prefers things to be more settled and orderly than that. But when you think about it, a painter contemplating a blank canvas or a writer looking at a blank sheet of paper relies on "Anything can happen" as the best and highest creative state. When you approach infinite potential as infinite creative possibilities, there's an opening to the kind of freedom "I" cannot experience within boundaries.

We are so conditioned to say "my" body and "my" mind that we automatically believe they are doing the acting and thinking we experience. But in metareality consciousness is doing the acting and thinking. We could—and should—shift into that perspective for one simple reason. The brain cannot make the leap into higher consciousness; only consciousness can. A violin string can't invent new forms of music, but the musical mind can use the violin as its instrument of physical expression. In the same way, the bodymind—an understanding that the body and mind are a single unit—is the physical expression of consciousness.

In music there are more possible combinations of notes than atoms in the universe, and yet we are comfortable with such vast possibilities. In chess, we play without anxiety in the face of 400 possible positions after each player has made one move, which explodes to over 288 billion after four moves. In everyday ways, like playing games, we take infinity for granted. It's all around us. Even if you had a limited vocabulary of two thousand words, which is what a typical five-year-old has, that's enough to create an endless string of word combinations, not to mention that nothing stops you from giving a common word multiple new meanings. (*Pet*, for example, is already a verb, a noun, and an adjective, as in "I pet my pet cat, who is the only pet I have"—and anyone is free to assign new meanings to the word. *Pet* could substitute for *coddled*, *luxurious*, *endearing*, and the like. It already has those connotations. New meanings can be invented out of the blue, the way *hip* and *cool* have wandered away from their literal meanings.)

Since we are so used to infinity in these simple guises, we can expand our comfort zone to be comfortable with infinity as a *personal* trait.

# Sudden Genius

"Anything can happen" is essential to being human. A leap of awareness can happen unexpectedly, as in the amazing phenomenon known as *sudden genius*, a term coined by Darold Treffert, a physician in Wisconsin who has become an expert in "exceptional brain performance." He defines *sudden genius* as a "spontaneous epiphany-like moment where the rules and intricacies of music, art, or mathematics, for example, are experienced and revealed, producing almost instantaneous giftedness."

Of the fourteen cases Treffert has studied to date, a striking example is a twenty-eight-year-old Israeli-born man he calls "K.A." As a performer, K.A. could casually pick out popular tunes on the piano one note at a time. One day he was in a shopping mall that had a piano on display. To quote his account of the next moment, "I suddenly realized what the major scale and minor scale were, what their chords were and where to put my fingers in order to play certain parts of the scale."

Without prior knowledge or ability, K.A. suddenly just *knew* how musical harmony worked. He verified this by searching for music theory on the internet, and, to his amazement, "[m]ost of what they had to teach I already knew." He was left baffled by how he could know a subject he had never studied.

*Such a phenomenon strongly supports a central feature of metareality—that human beings are already connected to infinite possibilities.*

We are kept from our hidden potential for all kinds of reasons—we've just seen how the ego edits and limits reality, for instance. Even when we are confronted by evidence to the contrary, the power of limitation prompts us to greet it with skepticism or overlook it entirely. The very fact that Treffert could get his observations on sudden genius published in *Scientific American* (July 25, 2018) is owed to a larger mystery, known as savant syndrome, which has a well-established history in medical and psychological practice.

Savant syndrome also deals with extraordinary abilities that defy explanation. There were already two different forms of this condition that Treffert is also an expert on. In "congenital savant syndrome," the extraordinary ability appears early in life. There are children who can tell you the day of the week that any date, past or future, falls on—so-called calendar savants—and others who can mentally generate prime numbers with the speed and accuracy of a computer. (A prime number is a numeral that is divisible only by 1 and itself. The sequence begins easily with 2, 3, 5, 7, and 11. But soon it becomes much harder to know the next number in the sequence without a calculator to help. For instance, 7,727 and 7,741 are both prime numbers, but none of the numbers in between them are.)

The other form is "acquired savant syndrome," where an ordinary person suddenly has astonishing abilities after a head injury, stroke, or other central nervous system incident. To these two versions Treffert has added "sudden savant syndrome" as a third type, as in K.A.'s story. None can be scientifically explained. In the congenital form, the child is often on the spectrum of autism or mental retardation. (In his celebrated 1985 book, *The Man Who Mistook His Wife for a Hat*, the neurologist Oliver Sacks recounts stories of "other-brained" people, including incidents of people with right-brain injury and autistic children who have extraordinary mathematical abilities.)

Savant syndrome had a somewhat alien reputation when it was confined to autism and brain injuries, but once sudden genius occurred in seemingly normal people, it came much closer to home. Treffert titled his article "Brain Gain: A Person Can Instantly Blossom into a Savant—and No One Knows Why." To a science-minded reader, *brain* is the most significant word in the title, because the standard explanation for all mental phenomena in our day is brain-centered. But it's not reasonable to say that a brain suddenly learned how music theory works without any external input, education, or training. (It's rather like saying that the brain knows all the cities in the world without ever having seen a map.) People who acquire sudden artistic ability typically become obsessed with painting and can hardly pull themselves away to do anything else. This is reminiscent of Picasso, who was obsessed with painting from early childhood, but it would be hard to imagine that people who had little or no interest in art became fascinated with making paintings overnight because their brain told them to.

For most people, art, music, and math are not critical aspects of life, and with only fourteen cases of sudden genius on record, so far at least, this strange phenomenon seems fairly exotic. But it turns out that sudden genius is yet another clue to the mystery of what it means to be human, and the clue has to do with infinity itself.

# Relating to Infinity

If someone has little or no interest in music and suddenly knows music theory without studying it, where did the knowledge come from? It wasn't sent through a teacher or textbook, which is the normal way of gaining knowledge. Somehow it was downloaded, we might say, from a source "out there" somewhere. But where? Music is a human creation—there are no chords or sonatas or symphonies in Nature. Nor is there an invisible library "out there," where all the information about music is stored. Besides, even if such a library existed, who would be sending people a download when they didn't want one?

We want to keep away from the head-spinning department, and metareality can feel bewildering. Metareality is the storage place for everything that has ever been thought and ever will be thought. Because infinity is boundless, metareality also contains *every possible* thought. Only recently have scholars of ancient Greece been able to decode the scraps of manuscripts that contain written music, and, with painstaking effort and guesswork, enabled the music to actually be played on reed pipes and drums exactly as Socrates might have heard street musicians as he wandered around Athens in the fourth century BCE.

To bring back ancient music is an act of re-creation. The mind is trying to remember what it never forgot, which sounds strange. That's not the same as retrieving information. To grasp what's really at stake, an image from everyday life might help: the cloud.

Anyone who accesses the internet has probably heard, if only vaguely, of the cloud, a place in cyberspace where the world's sum of information is held. The cloud has a capacity multiple times larger than the world's largest libraries. Every email is saved there, every online photo, every transaction at Amazon or search on Google. As universal as the cloud has become, few realize that it has a physical location.

Huge data centers have been erected to store all the information that was once stored on home and business computers. When you snap a photo of anything—a sunset, the Grand Canyon, a newborn baby—then crop it, improve the color, and send it to someone's smartphone, the image doesn't actually go from you to the other person without passing first through the cloud. Massive data centers in Loudon County, Virginia, each millions of square feet in size, house the actual cloud.

Accessing the cloud is instantaneous in cyberspace, but getting into one of these data centers is an arduous process. The core where thousands of powerful computers are located is protected by layers of security. A worker must pass through a retinal scanner that identifies the person (it even can tell that the person is alive, not someone contriving to use someone else's eyeball), then through a "mantrap," a room where the entry door must lock before the exit door will open, then through a squad of armed guards, and all along the way nine different passwords must be presented.

The fact that cyberspace is an actual place stands in stark contrast to consciousness. You don't go to a destination "out there" to remember the tune to Frank Sinatra's "Strangers in the Night," which comes complete with the sound of his voice and an image of his face. Metareality stores the whole experience, and we re-create it whenever we want to. Neuroscience says that we retrieve it from the brain, but the phenomenon of sudden genius casts shadows on that explanation, since people can retrieve knowledge they never had in the first place.

To go a step further, being human is creative, so we use consciousness to invent and discover new things. This isn't the same as re-creating ancient Greek music. But figuring out why some people are creative proves very difficult. Like the cloud, metareality has security measures in place. For a Leonardo da Vinci or a Ludwig van Beethoven, the riches of creation open easily and copiously. These artists are consumed by the creative act. At the other extreme, some people are completely shut off from artistic creativity. Creative geniuses, it seems, are hacking into metarcality. They relate to its infinite potential far beyond what we allow ourselves to do normally.

Efforts have been made to make creativity a skill, which sounds promising at first. It motivates a far-seeing company like Google or Apple to identify and harness the most creative minds. Let's go back to the book *Stealing Fire* mentioned above. In it Steven Kotler and Jamie Wheal go inside the making of "talent traps" across Silicon Valley. These enterprises, along with the revolution in social media via Facebook and Twitter, got their start through innovation, and it's only natural that, with billions of dollars at their disposal, Silicon Valley wants to spur ever more creative ideas—in a high-tech environment, survival depends on it.

These corporations would love to monetize the zone, but "aha" moments are just that, temporary flashes of insight. Sustaining insight and making innovation a way of life comprise the challenge. *Stealing Fire* describes how Google has created a work environment where one important element of creativity—flow—is engineered in ingenious ways.

The underlying notion is simple: if you want the mind to flow, make the workplace flow. Google takes this idea seriously and constantly tweaks its workplace so that it feels continuous with life outside work. There are bicycles for getting from one part of the campus to another, Wi-Fi on company buses, high-quality organic food, and a loose structure that allows ideas to move freely, rather than the traditional rigid hierarchy that seals management off from workers.

But these amenities have had limited success, because flow by itself doesn't define creativity. Kotler and Wheal focus on "altered states," as they term them, in which normal habits of mind drop away entirely. Whether induced by LSD or a total immersion tank, by meditation or sacred rituals, one can experience *ecstasis*, a word derived from Latin roots that mean "outside" and "to stand." An ecstatic state puts the person outside and beyond the mind's normal sense of self. In ecstasy we feel free, unbounded, blissful—and creative. *Stealing Fire* is essentially about how Silicon Valley seeks to monetize ecstasy.

Google encourages meditation for all its workers, for example, which gives access to a quieter, more expansive state of awareness. But despite knowing about the zone, flow, ecstasy, meditation, and witnessing awareness, it turns out that creativity is elusive. In 2013, searching for the means to solve "wicked" problems, a diverse group of bright minds met for the Hacking Creativity project, "the largest meta-analysis of creativity ever conducted." The guiding idea was that if creativity could be understood, anything would be possible.

To this end more than thirty thousand research papers on creativity were analyzed and hundreds of experts were interviewed, "from break dancers and circus performers to poets and rock stars." By 2016, two conclusions were reached.

First, creativity is essential for solving complex problems—the kinds we often face in a fast-paced world. Second, we have very little success in training people to be more creative. And there's a pretty simple explanation for this failure: we're trying to train a skill, but what we should be training is a state of mind.

Metareality, it turns out, is impossible to hack. Hacking is an intrusion, the cyber equivalent of breaking and entering. Creative people don't break and enter. What they do is much closer to evolution; something new is coming to light and taking physical form as a painting or a piece of music. To evolve is creative—the giraffe's long neck, the chameleon's ability to change color, and our own prize possession physically, the opposable thumb and forefinger, are the result of Nature's creativity. Working from the field of infinite possibilities, evolution proceeds one step at a time to create a whole creature, the same way you and I re-create the whole experience of hearing Sinatra sing "Strangers in the Night."

Every life form relates to metareality through evolution, tapping into its creative potential in an orderly, progressive way. The oddest creatures, as they look to us, aren't an assemblage of spare parts haphazardly cobbled together. They are expressions of a delicate adaptation that finds a perfect niche for itself apart from thousands of other life forms living in the same environment.

Every morning when it wakes up, the giant anteater (*Myrmecophaga tridactyla*) of Central and South America is hungry for ants. It is superbly adapted by physical evolution to seek out its insect prey buried deep underground. The giant anteater's front paws are equipped with four-inch-long, talon-like claws for ripping into an ant or termite colony. It has a two-foot-long, sticky tongue that it uses to slither around and ensnare the ants, which are then slurped back into its mouth.

One marvels at how specific these adaptations are. Even the giant anteater's jaws have been reduced to miniature size, since anything larger would be useless for burrowing into the narrow passages of anthills and termite colonies. But, at the same time, no human being would ever choose to be trapped at such an evolutionary dead end. To survive, the giant anteater must consume up to thirty thousand ants a day, and because its diet supplies so little nutrition, the animal doesn't have enough energy to do more than sleep for sixteen hours a day.

The beloved giant panda is also at a dead end. Although it is a true bear, and one of the earliest to evolve, the panda is badly served by having the wrong intestinal tract, which, unlike other bears' digestive systems, can only digest bamboo leaves. Why did the giant panda lose, or never gain, a typical bear's ability to eat almost anything? "Why" doesn't count in evolution. An adaptation occurs, and either it contributes to a species' survival or it doesn't.

*Homo sapiens* has eluded every dead end, including those that other primates ran into. Our nearest relations among living species are the chimpanzee and the gorilla, but we did not descend from them or any other ape, contrary to popular opinion. The last common ancestor between humans and chimpanzees lived around thirteen million years ago, according to the best current estimate, leading to a genetic split. One branch of the split evolved into chimps, gorillas, orangutans, and their relations, while the other branch led to our hominid ancestors.

A chimp is a modern species, and so are we. (Interestingly, the genetic evolution of chimps was two or three times more complex than ours, the result of two factors: first, chimps pass on twice as many random mutations from parent to offspring as human parents do—on average a human baby inherits around seventy new mutations from its parents. Second, there is something in genetics referred to as a "bottleneck," where not enough new genes are available. Fewer genes lead to fewer mutations, causing a species to remain closed off from new traits. According to genetic analyses, over millions of years chimps ran into three bottlenecks that squeezed their gene pool, while humans ran into only one bottleneck while migrating out of Africa 200,000 years ago. Once out of the bottleneck, our species expanded explosively to cover the globe.)

The great reality shift poses a profound mystery that cannot be unraveled by examining our genes. Even with twice the mutation rate as *Homo sapiens*, chimpanzees have not gained self-awareness. That's not the same as lacking intelligence. Primate studies increasingly show that chimpanzees are much closer to human intelligence than was previously thought. According to primatologist Frans de Waal, the idea that only humans can make tools is now "an unsustainable position. Then we also got the apes-have-no-theory-of-mind claims, which now have been seriously weakened. The culture claims, the idea that only humans are great at cooperation, and so on—none of [them] really holds up."

*Theory of mind* is a term from philosophy and psychology about knowing someone else's mental state without being told. A shorthand definition is "mind-reading." Many dog lovers would swear that their pet knows when the owner is happy, sad, or angry. But proving that any animal understands that it has a mind remains controversial. It is fascinating to consider that our common ancestor thirteen million years ago had the hidden potential for traits that flowered in chimps along one branch of the divide and in humans on the other.

But chimps didn't experience the great reality shift that gave *Homo sapiens* self-awareness. There are limits to what a chimp can learn and understand. For example, if you place a peanut under a yellow cup and no peanut under a red cup, a chimp will quickly learn the difference and always pick the yellow cup. But if you present a chimp with two weights and give him a reward if he picks up the one that weighs less, he won't make the connection and will keep picking up the two weights at random. Likewise, if one chimp teaches herself how to unlatch a complicated box to get at some food, another chimp cannot learn how to solve the problem simply by watching the first chimp.

These differences between us and our closest relatives go only so far. Chimps and gorillas have a much wider range of possibilities open to them than a giant anteater. But they aren't remotely close to us in accessing infinite possibilities. The great reality shift transcends genetics, and no matter how much we see ourselves when observing other primates, they don't see themselves in us, because they simply can't. Some evolutionary dead ends lead to lives as complex as the ones led by higher primates, but they are dead ends nevertheless.

The miracle of being human is that we evolved in multiple dimensions. Everything about us—behavior, abstract thinking, curiosity, individual personalities, social networks—exploded beyond any precedents, as if life on Earth rushed into unknown levels of aspiration. A master of three-dimensional chess would be dumbfounded to lay eyes on our game board, which has numerous levels and adds new ones all the time. But this master wouldn't be able to see our setup, because it exists and evolves in consciousness. This is such an important point that we need to go into it more deeply.

## Awareness in Multiple Dimensions

Infinity can stay in one dimension, or it can exist in several dimensions at once. You might think of a number like two-thirds spinning out a single string of digits—0.6666666—to infinity, like a mathematical silkworm spinning an endless thread. That's an example of infinity occupying one dimension. But flow, like creativity, is a state of awareness that covers multiple dimensions. A person is "in the flow" when everything seems to go right, obstacles melt away, and answers come effortlessly. Each aspect occupies its own dimension, yet somehow they get coordinated when flow is experienced. There is a sense of calmness and also the experience of happiness, sometimes to the point of ecstasy. If the flow is powerful enough, it can be all-consuming, giving the impression that the creative ideas are coming of their own accord, using us as a vehicle, the way a playwright uses actors to mouth his words.

Flow feels desirable but quite mysterious, since it comes and goes, and some people never experience it. The mystery is solved when you realize that flow is the unobstructed access to metareality. In it we are experiencing wholeness. Puddles don't flow, but the ocean does. Only wholeness can orchestrate a state of awareness that embraces multiple dimensions. Our hominid ancestors undoubtedly became multidimensional long, long ago. There was a mass awakening that left its traces in the first known cave paintings. These were only made possible by a seismic shift in consciousness. The accepted date by scientists for the first paintings of animals on cave walls has been moving back further and further. The record was long held by Europe, first in a complex of caves in Lascaux, France (about 17,000 years ago), then by separate caves in France and Romania (30,000–32,000 years ago), representing a huge leap back in time. Now the current record is held by a cave system in Indonesia on the South Sulawesi Peninsula, still in use by the native people. Radiocarbon analysis has dated the Indonesian paintings at between 35,000 and 42,000 years ago.

Gazing at photos of these prehistoric artifacts (easily found online) is a form of time travel, not physical but mental. At the oldest French site, the Chauvet-Pont-d'Arc Cave, which is on a limestone cliff above an ancient riverbed, now dry, the paintings are nothing less than art. Discovered in 1994 and soon named a World Heritage Site by UNESCO, the Chauvet Cave features very large, very well-preserved animal figures by the hundreds, depicting thirteen species. In other Ice Age paintings, the chief subjects are herbivores, like horses and cattle, but the Chauvet artists added predators, like the cave lion, the panther, the bear, and the hyena.

The circumstances behind the art were arduous. The cave walls are deep inside, far from outside light, which means that the painters worked in the dark by the light of flickering torches. As the light bobbed back and forth, the painters still had such a sure hand that the major curves of each specific animal—horns, back, head, and legs—were done in a single stroke, the way an accomplished draftsman would do them today. The walls were apparently cleaned and smoothed to give a suitable pale area, a bare canvas, before the actual painting began. Two woolly rhinoceroses are pictured butting heads, reflecting a desire by the original artist to portray scenes, and many animals are shown in motion, not like a child's stick-figure drawings.

As you could imagine, the desire to paint a picture isn't simple and never has been. Are these paintings mystical, ritualistic, or magical? The mental state of the artists cannot be conclusively determined. There is beauty in how the animals were drawn, or are we simply imposing "beauty" from our own familiar concepts? Whatever cave painters had in mind, it went viral, so to speak. The same images are repeated by cultures thousands of miles apart.

A smaller mystery within the larger one is why the cave painters didn't depict themselves. At Chauvet thirty thousand years ago, the painters portrayed no full human figures and only one partial figure (the lower half of a female body, with prominent sexual organs), but there are numerous outlines of human hands imprinted on the wall. These were done by blowing red ochre pigment around someone's hand as it was pressed against the stone. Outlines of hands also appear in the Indonesian caves thousands of years earlier, then in Argentina between nine thousand and thirteen thousand years ago, and in petroglyphs of the Anasazi, or ancient ones, of the Southwest of the United States. These scattered handprints lead to no definite conclusion. Perhaps they mark tribal allegiances, or simply the message *I was here.*

The explosion of sophisticated art, on a global scale, testifies to a hidden potential coming to life as an awakening in consciousness, vibrant and full-blown. Art must be considered evolutionary in *Homo sapiens,* because it has persisted as a dominant trait of our species in every society without dying out. Yet it seems unlikely that art somehow provided a survival value. Cave painting testifies to prehistoric humans' caring for something beyond physical needs, and this something emerged directly from their consciousness.

To me the Chauvet cave paintings clearly drew on the whole mind. A prehistoric Picasso didn't step up to the blank canvas of a limestone wall and let fly his genius. *Homo sapiens* evolved much more holistically. At the very least, the desire to create art involves the following attributes we can all identify with, even if we aren't artists:

Curiosity
Intelligence
Purpose
Motivation
Diligence
Eye-hand coordination
Learning a skill set

These mental traits have to exist in order to build the Taj Mahal, invent the internal combustion engine, or paint a woolly rhinoceros on a cave wall. And the whole collection of traits has to come together into a single intention. Figuring out how this happens may be impossible scientifically. There is no way to gather evidence when what you are looking for is invisible. Even so, the capacity to use the whole mind convinces me that our remote ancestors contained the seeds of metahuman. Being able to access the whole mind destroys the myth that our ancestors were primitive. Their potential was already infinite.

. . . . . . . . . .

# A Brain Teaser

Once *Homo sapiens* found a path to whole mind, the experience became embedded in human evolution. If we examine ourselves closely, it is clear that we are whole-mind creatures. Let's test this proposition with a simple chore like driving to the grocery store to buy a child's birthday cake. Behind the simplicity of the task is a complex of mental activities we take for granted. The ingredients in the mix include the following:

Knowing the concept of "birthday cake" and matching the words
   to a physical object.
Wanting to do something nice for your child.
Putting the cake on your to-do list.
Knowing how to drive.
Remembering the route to the grocery store.
Giving priority to this task over other demands on your time.
Selecting one cake out of many, using judgment about which one
   your child will like the best.

These things span emotion, intention, visual recognition, memory, motor coordination, and stored skills—for one small errand! Neuroscience can locate and isolate where some of these mental activities are represented in the brain, but how they mesh into a unified outcome—shopping for a birthday cake—cannot be explained. The process of getting billions of brain cells to fire in sync makes herding cats look simple. It's more like wrangling every cat in the world.

Even less can anyone explain how the brain shifts gears without any visible transition from A to B. One incredibly complicated pattern of brain activity spontaneously shifts to another. Unlike a car, there's no transmission or gearshift. Imagine that you are reading a novel and forget that there's a roast in the oven. Absorbed in *The Lord of the Rings* or *Jane Eyre*, you suddenly notice the acrid smell of smoke in the air. Instantly, this simple sensation of smell causes you to leap into action, and the book is forgotten. Carbon molecules stimulating receptors for detecting odor throw your brain into a new pattern of coordinated action. Relaxation turns to stress in an instant.

To say that your brain noticed the burning roast doesn't explain your reaction. Your brain processed the raw data of smoke particles entering the nasal passage, but it takes a person to notice the smell and give it significance. The same smell, when applied to charring a steak on the grill in summer, provokes no alarm, even though the physical process taking place in the nose and brain is identical. Electrochemical activity buzzing around the cortex forms one pattern for the house burning down and another for a barbecue. That's all a neuroscientist has to work with, and it's not nearly enough.

Even if neuroscience could map every minute area of brain activity, down to the smallest detail—as is currently being attempted by the National Institutes of Health through the federally funded Brain Initiative—consciousness would be undetectable. If the brain is a red herring in regard to a whiff of smoke, it may be a red herring in lots of other ways as well. Two would-be poets could be writing a sonnet, causing specific brain activity to light up on an fMRI scan, but nobody would be able to pick out which writer is Shakespeare and which a duffer. Einstein's brain was autopsied after death to determine if the world's greatest genius, as he was popularly tagged, had an unusual brain; he didn't. On YouTube, you can find numerous videos of child prodigies playing the piano at the age of three or four when, according to normal brain development, such a complex display of muscle coordination shouldn't be possible. Attaining skill at piano playing ordinarily takes years or decades to achieve.

Having escaped the constraints of physical evolution, *Homo sapiens* stopped being puppets of the brain. Despite every finding revealed by neuroscience—many of them fascinating and important—the bottom line is that you rule your brain, not the other way around. The path to metahuman is a path of self-awareness about this very thing. You can control who you are, even if you think you can't. The trail of clues has led us far along the path. Now we are ready to undertake a reality shift even more radical than the mass awakening of our remote ancestors. In the next reality shift, we will define ourselves as beings endowed with infinite possibilities. Metareality becomes home when everyone agrees that we belong there.

# 4

---

# METAREALITY OFFERS
# ABSOLUTE FREEDOM

Freedom is the opposite of feeling trapped. We've created virtual reality on such a huge scale that people can go about their lives not feeling trapped. But, in truth, the whole setup is a trap. Hamlet was trapped in a dilemma over murdering the man who murdered his father. Indecision would lead him either to murder or to suicide—eventually it leads him to both—and at one point Hamlet speaks to his friends Rosencrantz and Guildenstern with all the symptoms of clinical depression:

> I have . . . lost all my mirth, forgone all custom of exercises, and indeed it goes so heavily with my disposition that this goodly frame, the earth, seems to me a sterile promontory; this most excellent canopy, the air—look you, this brave o'er-hanging firmament, this majestical roof fretted with golden fire—why, it appears no other thing to me than a foul and pestilent congregation of vapors.

No actual person burdened by depression speaks so beautifully, no doubt, but Hamlet still harbors the great optimism about human potential that fueled the Renaissance. Immediately he adds:

> What a piece of work is a man! How noble in reason, how infinite in faculty! In form and moving how express and admirable! In action how like an angel, in apprehension how like a god! The beauty of the world! The paragon of animals!

Four hundred years later, we are the same species, but optimism has become a struggle. Looking around, would you agree with Shakespeare's praise of human beings as "The beauty of the world! The paragon of animals!"? I've been arguing that *Homo sapiens* is the only species on Earth that is multidimensional. This seems like the greatest possible gift any life form could receive. It gave our minds an open ticket to imagine anything we want. It saved us from the evolutionary dead end that befell the giant anteater, the panda, and every other creature that is trapped by physical evolution. It is easy to become intoxicated by the prospect of our own boundless potential.

So it is strange but true that human beings don't accept this gift of unlimited potential. We are starkly divided—one side of human nature craving freedom, the other side deeply fearing it. By *freedom* I mean much more than not being in prison or under the thumb of the powerful. Freedom is boundless awareness, the very definition of *metahuman*. When anything is possible, we are most ourselves. When only some things are possible, we are all too human.

Freedom has no fixed rules. It is without walls, and any thought is permissible. No matter how good your life is, you are not living in freedom, unfortunately. You were born into a reality hemmed in by rules, walls, and thoughts you are not supposed to think. In these pages we have been dealing in forbidden thoughts, the kind that undermine virtual reality and expose its illusions. Why? To build a new reality based on freedom.

The ultimate forbidden thought is "None of this is real." It wipes the slate clean of all illusions. What makes such a thought forbidden isn't that the police will knock on your door and punish you for breaking the law. No one will even know it if you opt out of virtual reality.

What makes it forbidden to think "None of this is real" is something personal: the fear of what comes next. Absolute freedom is terrifying. It expands the unknown as far as the eye can see. That's the main reason that human history isn't about absolute freedom; it is about testing the next boundary, and then moving beyond it to test a new boundary. From the perspective of metareality, *Homo sapiens* has never needed to respect so many boundaries. A history of "thou shalt not" has unfolded with so much fear behind it, and such powerful enforcement of the rules, that our evolution has had to struggle every step of the way.

The root cause of living inside boundaries is the divided self. We have defined human nature simultaneously as something to celebrate and something to fear. Getting beyond the divided self is crucial, because, like every part of virtual reality, the divided self has no validity other than what we give it. The war between good and evil, light and darkness, creation and destruction, is a mental construct. It was born out of our deep confusion over who we are. When a species has an open ticket in terms of its future evolution, the destination printed on the ticket is Everywhere. *Homo sapiens* still holds this open ticket in hand, but we plot the course of history by ups and downs, progress and setbacks, tyrants and liberators, war and peace. Every day we force reality to fit the model of the divided self.

Testing boundaries is a stage of development. Every child tests the rules laid down by his parents and grows up with an inner map of dos and don'ts that, for the most part, lasts for life. But it would be far better if our inner map showed us how to use our ticket to Everywhere. Fortunately, metareality gets closer the more we realize who we are. It has been getting closer for tens of thousands of years. To see the invisible evolution of consciousness requires us to look at our species with new eyes.

# The Iceman's Tale

Although we have no exact date for it, human beings couldn't explore multiple dimensions until they became aware that this was possible. Prehistoric humans made that breakthrough and, like the other invisible things in our past, we can read the minds of our remote ancestors by investigating a handful of physical clues. What these clues reveal is fascinating. While warring between the darkness and the light inside us, we were also expanding into freedom. Limitation pulled us one way, freedom the other. This tug-of-war will take some explaining.

In 1991 some hikers chanced upon a natural mummy preserved in glacial ice at an altitude of approximately 10,000 feet on the side of an Alp on the Austrian-Italian border. Named Ötzi by anthropologists (after the mountains he was discovered in) and "the Iceman" in the popular press, this male from 3500 BCE is a frozen snapshot of life in Europe during the Chalcolithic Period, also known as the Copper Age—*chalcolithic* is derived from the Greek word for copper. By this time prehistoric humans had discovered how to smelt copper ore, but had yet to mix it with tin to form bronze, a harder metal.

If you were hiking the South Tyrol Mountains and met the Iceman two hours before his death, you'd have seen a thin, almost malnourished male about five feet, two inches in height, perhaps on a hunting trip, sitting on a rock and having a meal. He was carrying bread made of wheat and barley; there were also fruits and roots in his supplies, and the dried meat of chamois, red deer, and ibex. If he smiled at you, Ötzi would have shown teeth riddled with cavities, probably because his diet was high in carbohydrates.

The mummified Ötzi's exposed skin displays numerous tattoos, sixty-one in all, mostly horizontal lines and crosses, primarily running up and down the spine and behind the knees. According to the anthropologists who studied his remains, he had been sick three times in the previous six months before he died. For a forty-five-year-old, Ötzi was in bad shape by modern standards, already showing wear and tear in the knees and ankles, and osteoporosis of the spine. (The tattoos, made from powdered black soot, might have been some form of ancient medical treatment. There's an ancestral tradition in every culture of tattooing to relieve pain, and the Iceman's tattoos are clustered around the joints and lower back, where he would certainly have suffered from chronic pain.) Unlike others who settled in agricultural communities, this man took long walks all his life, which adds to the speculation that he was perhaps a shepherd or a hunter.

No other early European has been studied as thoroughly as the Iceman, down to the contents of his stomach, the minerals in his hair (showing traces of copper and arsenic), and the exact composition of his DNA. His genes reveal that he had brown eyes and dark hair, along with a predisposition to heart disease. But my mind goes to the invisible inner journey that *Homo sapiens* had taken by the Copper Age, strongly hinted at in the physical findings.

There had been many awakenings in consciousness to make Ötzi possible. We have to imagine what these awakenings were like. Agriculture, sewing, the tanning of hides, the discovery that rocks could be heated until they oozed with pure metal—these are titanic leaps in awareness. The Iceman's brain was outfitted to perform extremely sophisticated tasks.

Keeping thoughts in consecutive order, each leading to its own conclusion, isn't a modern trait. A highly organized mind made it possible for the Iceman to navigate a surprisingly complicated existence. We can infer this, for example, from his elaborate costume: shoes, a hat, a belt, leggings, and a loincloth, all fabricated from different leathers. His cape was of woven grass, pointing the way to the intricate weaving that lay ahead as cultures evolved.

The Iceman feels very far away from us and yet very close at the same time. However, as members of *Homo sapiens*, we inhabit the same species of consciousness as Ötzi. Some of the strongest evidence of our kinship across the centuries is tragic. On the day of his death, the Iceman was heavily armed with an ax, fashioned out of 99 percent pure copper, and a flint knife, both crafted with wooden handles. Over his shoulder was slung a longbow and a quiver filled with fourteen arrows. Clearly, he met adversaries similarly armed, because an arrowhead had pierced his cloak at the shoulder and lodged itself. The arrow shaft was broken off, probably in an unsuccessful attempt to remove the arrowhead. There was DNA on the body from three other individuals: his friends, his enemies, perhaps someone he killed before he himself was slain?

Ötzi was found facedown, suggesting that someone else placed him that way while trying to pull the arrow out of his back. But Ötzi quickly bled to death and was covered by ice from an ice storm almost immediately afterward. It's not hard to imagine a link from him to familiar emotional states in modern life: tribal warfare and loyalty, angry attacks and defense, the desire to save a fallen comrade. The war between light and darkness had already become endemic in human society.

# The Lost Source

The Iceman wasn't the first person to die by violence. Skeletons of females from fourteen thousand years ago in Mexico display broken bones and other wounds, indicating that they had suffered physical injuries consistent with abuse, rather than accidents. Somewhere between metareality and human reality, bad things happen. For millennia, the blame has been ascribed to sin or human failing in general. I'm ascribing it to the divided self, where peace and violence were given separate compartments. Projected outward, they became institutions of war—armies, weapons makers, armorers—and institutions of peace—law courts, codes of justice, religions that depicted a God of mercy.

Societies learned to live with the contradiction. Rome saw itself as the great harbinger of peace in the world and, yet, in conquering the territories of modern-day Spain, Britain, and Germany, Julius Caesar led campaigns of unspeakable barbarity. In one village he ordered the hands of every adult male cut off; all told, his conquest of Gaul took as many as two million lives. The Romans celebrated Caesar (offering him the title of emperor before he was assassinated in the Senate), but a popular saying exposes a deeper truth: *Homo homini lupus*, "Man is wolf to man." We have not seemingly evolved past this brutal wisdom, since the United States is both a global peacekeeper and the world's largest arms dealer.

Being trapped in the divided self isn't the whole story, however. If we examine the Iceman with new eyes, from the perspective of consciousness, we recognize that something immense was happening in human history. Anthropologists refer to the "cognitive explosion" or "intelligence explosion" that hurtled Stone Age humans forward. One cannot deny the evidence offered by the complexity of the Iceman's tools, weapons, diet, and clothing. It took new and ingenious thinking for early humans to arrive at such sophisticated mental activity.

To explain the cognitive explosion, various arguments point to the higher brain and genetics, or milestones like the discovery of fire, which brought early humans closer together, making collective thinking possible. As usual, scientists rely on the physical evidence. However, I think the real story lies in the expansion of awareness, which occurred before the physical evidence appeared. Consider what kind of awareness it took to make the first bow and arrow. Given nothing but a forest and a single sharp stone blade, could you invent the bow and arrow?

A startling series of YouTube videos, titled "Primitive Technology," shows how it might be done. A single barefoot man in shorts cuts down a small tree by hacking at it with a chunky rock, sharpened on one side into a blade. He splits the wood using the same blade until he has a three-foot-long piece that he tapers at each end until it is limber—this is the bow, which gets notched at both ends to hold the string. The string comes from peeled green bark from a sapling. Allowing the sapling bark to dry and twisting it twice, the string is attached to the bow. For arrows, the man uses a thin sapling, and by peeling away the bark and scraping down the wood, he fashions a long, thin shaft.

At this point our primitive craftsman needs feathers to make the arrow fly straight; these are the only artifacts not related to a single stone blade. Using the back feathers of a chicken, the craftsman narrows them down, then smooths the edges with the tip of a stick heated red hot in a fire. Having assembled bow and arrow (along with a quiver made of bark to carry the arrows), the craftsman proved the viability of his weapon by firing an arrow into a six-inch tree ten yards away with deadly accuracy.

It's hard to describe how astonishing this process seems when first viewed, but, on reflection, a modern craftsman duplicating the efforts of a hunter over forty thousand years ago is cheating. He already knows what a bow and arrow look like and what they're for. The first maker of a bow and arrow worked with pure ingenuity and a sense of discovery. What did it take to invent the first bow and arrow? It took expanded awareness, and, in particular, it took some features of awareness possessed only by *Homo sapiens*, as follows: The bow-and-arrow maker had to conceive of what he wanted to do and then figure out how to do it. He no doubt experimented and tested various options, just as Thomas Edison tested different substances for the filament of the electric light bulb before he hit upon tungsten. Moreover, as he began to craft his weapon, the bow-and-arrow maker had to focus his attention and keep his mind on what he was doing. If distracted, he had to remember his task and return to it.

Each of these mental faculties—attention, intention, focus, curiosity, diligence—are not thoughts. They are the foundation of thoughts, like invisible brick and mortar. Without them, nothing can be fabricated (an alternative name for our species is *Homo faber*, man the maker). Learning what consciousness was capable of lies at the heart of human evolution. The multiple dimensions we grew into are dimensions in awareness, which aren't the same as physical dimensions in space.

This is the right place to specify the difference between consciousness and mind. It needn't be abstract. Mind is the activity of thought. Consciousness is the field of pure awareness. In an analogy from ancient India, consciousness is the ocean, mind is the waves that play across the ocean's surface. Once you grasp the difference, a radical insight dawns. Mind only came about because consciousness started to move within itself. The field of pure awareness began to vibrate, and from those vibrations, consciousness acquired the familiar mental attainments that it took to make a bow and arrow (and everything else in the field of technology for the next forty millennia).

Eventually, human beings got lost in the complexity, bewildered and yet fascinated by everything the mind can do. That includes the darker side, for we have inherited the mind's capacity to make us violent, afraid, depressed, sad, and conflicted—all the traits of the divided self. Living with the divided self has been the source of profound sorrow and sometimes a deep wish to start all over again. There's a reason that every culture has a myth about a golden age or a lost Garden of Eden. We are disappointed in how we turned out, and the tug of nostalgia leads us back to a pristine time when we were more innocent and better.

If we drop the mythic trappings, however, what humans lost was the connection to our source—or pure consciousness. But we have a path back to Eden that isn't mythical. Somewhere in the evolution of the human mind, at a time unreachable from the present, *Homo sapiens* became self-aware. Was this a momentous shift or a gradual development? No one knows. Early humans started to go beyond everyday thinking, which is largely utilitarian, devoted to the busyness of daily duties and demands. Going beyond became solidified, we might say, in God, a superhuman fashioned in our own image. At other times, however, going beyond meant contemplating the nature of awareness. We can call this the awareness of being aware. We thereby gained access to our source. Only when you *know* that you are conscious can you begin to explore where consciousness comes from.

# Saved—and Cursed—by Reason

Throughout history, this search for consciousness has been haphazard and hazardous at the same time. Being human encompasses a culture of nonviolence in Buddhism and early Christianity, but also warrior cultures like the Mongols and the Vikings. Traits we take for granted, like romantic love, didn't exist in many early societies in the Mediterranean Basin, including the Greeks and the Egyptians; chivalry toward women was unknown outside noble circles until the rise of courtly knighthood in the Middle Ages. Children were not considered exemplars of innocence as a matter of course—in medieval Christian doctrine, sin arrived at birth, and under English common law a child was treated as a father's property, or chattel. Human dignity itself was betrayed by the practice of slavery.

What this dismal history indicates, if we look for a common denominator, is that the human mind knew itself in fragments, which accumulated into shared stories built up over time. For example, the first life account of the Buddha, who lived in the sixth century BCE, wasn't written until around four hundred years after his death, and it was very brief, being included in an account of the twenty-five Buddhas who preceded Gautama Buddha. The first true biography was an epic poem, dating to the second century BCE, written by a monk called Ashvaghosha, under the title of *Buddhacarita* (*Act of the Buddha*). Understandably, myths and miracles are interspersed with facts that may be reliable accounts—the important thing was to amass from many fragments a saintly life suitable for worship and veneration. Likewise, the New Testament comes from many sources—principally, it is thought, the various early churches scattered throughout the Roman Empire—and very big differences crop up between the four Gospels.

All the stories that became woven into civilization were created collectively, even when it is traditional to refer to a single author like Homer or a Gospel writer. The original texts always underwent alterations later. Some stories became inspiring. Others became part of a people's identity or their way of worship. Against these positive effects, every story has severely limited our infinite potential. Metareality has no story because it is outside time and therefore outside history. Unlike the mind, which leaves a trail of events behind it that historians can study, consciousness has no beginning or end.

At present the dominant story, collectively speaking, is scientific, and one aspect of the mind is given credit for advancing human evolution: rational thought. If we pity our forebears for their difficulty in getting past superstition and myth, the future may pity us for glorifying the rational mind and neglecting the whole mind.

In his 2018 book, *Enlightenment Now*, the widely read Harvard psychologist Steven Pinker takes 450 pages to fully extol the triumph of reason in recent cultural history. Pinker makes "the case for reason, science, humanism, and progress" (the book's subtitle), much as it would have been made in France during the eighteenth-century Age of Enlightenment. In fact, he looks to that period as the turning point in Western civilization.

For the vast majority of people in the West, Pinker is preaching to the choir when he characterizes *enlightenment* as secular, free-thinking, rational, and dedicated to progress. I know this in my role as an MD trained in scientific medicine. Rationality has few greater triumphs, and the rise of modern medicine was part of a much larger campaign to wipe out the scourge of ignorant superstition.

As a prime example, no one in the fourteenth century had a rational explanation for the horrors of the Black Death, which killed perhaps a third of Europe's population, around twenty million people, between 1347 and 1352. Supernatural explanations became widespread, and there was a rash of persecutions of Jews and witches. Three centuries later, you would think society would have learned its lessons and realized how superstitious they had been. No such luck. When William Harvey in the seventeenth century scientifically confirmed that the heart pumped blood to every part of the body and back again, many still believed in witches. It is estimated that more were executed in the century after Shakespeare's death in 1616 than in the previous century.

Harvey himself visited women suspected of witchcraft and became a prominent and skeptical opponent of such superstitious thinking. In one of these visits, he took a toad, which one woman thought was her familiar sent by the Devil, and dissected it before her eyes to prove that nothing supernatural existed in the creature.

Pinker praises the rise of enlightened reason with fervor and confidence:

> What is enlightenment? In a 1784 essay with that question as its title, Immanuel Kant answered that it consists of "humankind's emergence from its self-incurred immaturity," its "lazy and cowardly" submission to "dogmas and formulas" of religious or political authority. Enlightenment's motto, he proclaimed, is "Dare to understand!"

This is such a widely accepted view of enlightenment that there's little that's daring about it. A secular, scientific culture not only takes pride in rationality, it worships rationality with a faith as absolute as the religious dogmas Kant wanted humanity to escape. What would be daring is to find a way to escape the unfounded exaltation of reason. For rationality has wreaked as much havoc as unreason—medieval monks, wedded to the "lazy and cowardly" dogmas of Catholicism, as Pinker describes them, didn't drop the atomic bomb, invent chemical warfare and biological weapons, or despoil the environment to the point of humanity's self-destruction.

Reason has certainly created many good things, but in Pandora's box there was diabolical creativity, too. When freedom of thought became widespread, there was nothing to stop diabolical creativity from devising newer means of mechanized death, which has proceeded apace since the Roman catapult and the crossbow in the Middle Ages. Reason hasn't been able to curb its own creations, pursued in every case with the backing of rational explanations for why horrific weapons were necessary to develop.

Bypassing this fatal flaw, Pinker's optimism paints a picture of progress on many fronts. The core of *Enlightenment Now* is a set of seventy-five charts illustrating the advance of humankind up to this moment in history, countering the popular notion that the world is falling apart. The range of topics illustrated by these charts covers everything from the tone of the news people read, life expectancy, and child mortality to malnourishment and famine deaths; from Gross World Product and global income distribution to American leisure time and the cost of jet travel.

It's an impressive summing up of what rationality and progress have achieved. But Pinker's overview leaves consciousness for very late in the game. The word is barely present in the book, and the main references, beginning on page 425, are skeptical and debunking. He raises the issue of the "hard problem," a phrase coined by philosopher David Chalmers, which poses the riddle of why human beings have subjectivity—in other words, the world "in here."

True to his faith in reason, logic, and science, Pinker says that what goes on in the subjective domain (i.e., thoughts, sensations, images, and emotions), along with the behavior we exhibit based on subjective events, are "obvious Darwinian adaptations. With advances in evolutionary psychology more and more of our conscious experiences are being explained in this way, including our intellectual obsessions, moral emotions, and aesthetic reactions."

In other words, the mind can be logically explained—or explained away—by the same Darwinian drive that shaped every living creature: survival of the fittest. The possibility that *Homo sapiens* is a consciousness-based species isn't touched on, and one imagines Pinker would laugh it away. He isn't someone I want to single out for special criticism—

Pinker's argument is part of a much larger trend in science—but saying that humankind feels the difference between right and wrong (a moral emotion), craves the truth (an intellectual obsession), and loves beauty (an aesthetic reaction) because those things were survival traits is seriously misguided. By refusing even to take consciousness seriously, Pinker returns us to being higher mammals who happened to become conscious in order to get an advantage in finding food and gaining mating rights.

Some pure Darwinians stop there, but Pinker realizes that consciousness needs a slightly more credible explanation, because the hard problem doesn't ask how humans became conscious; it asks what consciousness is. Again not singling Pinker out for blame, he joins many other "apostles of scientism," as one of his critics called them, by saying two things: (1) Consciousness is probably an illusion created by complex brain activity, and (2) the whole issue is basically irrelevant.*

To hold that the entire subjective world is an illusion shows how blind the dogma of rationality has become. Citing the philosopher Daniel Dennett, one of the strongest consciousness deniers, Pinker is impressed by Dennett's view that "there is no hard problem of consciousness: it is a confusion arising from the bad habit of a *homunculus* seated in a theater inside the skull. This is the disembodied experiencer." The word *homunculus* means a small human being, and it has various meanings in science and philosophy. Supporters of the view that consciousness is an illusion like to say that people mistakenly believe there is a "little me," or an individual self, inside us. Dennett dismisses the self as an illusory artifact of the whirling, teeming brain. Pinker can't bring himself to go that far, but many neuroscientists do. In a sense they must, because if the brain creates the mind, then the notion of the self can't be anything but another product of brain activity. If every self is a delusion, then why should we

---

* For an extended discussion of the hard problem, see *You Are the Universe*, coauthored by me and the widely published physicist Menas Kafatos.

believe a scientist (i.e., a self) when he tries to debunk the self? Wouldn't we be taking the word of an illusion? The whole argument is hopelessly illogical.

# Behind the Mask

Many people consider science a liberating force, and it is. But it has also served to block the way to the absolute freedom that metahuman holds out. Claiming that consciousness is an illusion or irrelevant or just Darwinian evolution at work in the brain shows how far working scientists of integrity and intelligence are willing to go to adopt the role of deniers. The deepest irony is that quantum physics, which peered behind the mask of the physical world a century ago, gave rise to generations of physicists who feel that consciousness isn't worth bothering about.

There have been moments of going back to the future, though. The eminent Russian-American physicist Andrej Linde made important contributions to "inflation theory," which deals with the first stages of the infant universe when it was smaller than the period at the end of a sentence. Later on, he became an early supporter of the "multiverse," which posits that there are innumerable other universes beyond ours—the current estimate runs to $10^{500}$ different universes, or 10 followed by 500 zeroes.

In a 1998 paper titled, with characteristically sweeping ambition, "Universe, Life, Consciousness," Linde not so subtly sabotages the accepted worldview of modern physics. "According to standard materialistic doctrine," he asserts, "consciousness . . . plays a secondary, subservient role, being considered just a function of matter and a tool for the description of the truly existing material world."

One envisions heads nodding in the audience as Linde says this, with no hint of the contrarian thinking that is to come. He continues:

But let us remember that our knowledge of the world begins not with matter but with perceptions. I know for sure that my pain exists, my "green" exists, and my "sweet" exists. I do not need any proof of their existence, because these events are a part of me; everything else is a theory.

This last statement, "everything else is a theory," indicates that something strange is afoot, and the radical nature of that something begins to build. Linde notes that in the scientific model, perceptions are subject to the laws of Nature, like everything else in the universe:

This model of [a] material world obeying laws of physics is so successful that soon we forget about our starting point and say that matter is the only reality, and perceptions are only helpful for its description. . . . But in fact we are substituting [the] reality of our feelings by a successfully working theory of an independently existing material world.

Linde's audience was probably squirming by this point, because he had spotted that the world "out there" is a substitute for perceptions, which rightfully should be considered the starting point for any model of reality. After all, as he declares, science isn't needed to prove that green is green and sweet is sweet. It can't offer such proof; only subjective experience validates the most basic perceptions.

Only perception, without a doubt, is real. In other words, Linde was turning the tables on a worldview that puts matter first and consciousness second. He goes on, "And the [materialistic] theory is so successful that we almost never think about its limitations until we must address some really deep issues, which do not fit into our model of reality."

Linde's version of a deep issue would focus on the precise nature of how creation came about in terms of big bangs, other universes, subatomic particles, and so on. But when he speculates that consciousness is unavoidable in reaching an answer—a realization the quantum pioneers had reached long ago—this is likely to be a bridge too far. Science today is at its most materialistic, thanks to the surging success of technology. Linde breaks ranks, but he still remains locked in the technicalities of a professional physicist. Yet he has joined Stephen Hawking and others in realizing that science, however advanced, does not describe reality. Linde goes even further. To quote from an online article, "Consciousness and the New Paradigm," by Adrian David Nelson,

> Linde ... has urged his colleagues to remain open-minded toward a fundamental place for consciousness in quantum mechanics. "Avoiding the concept of consciousness in quantum cosmology," he warns, "may lead to an artificial narrowing of our outlook."

If you unpack Linde's seemingly mild warning, he's implying that the universe cannot exist without consciousness. Nelson goes on,

> Linde is also one of several respected physicists who've pointed out that the quantum wave function of the entire universe could not evolve in time without the introduction of a relative observer.

In a word, the universe since the big bang could not have expanded, leading in one small corner to life on Earth, without someone to see it happen (the "relative observer"). So who is this mystery observer? The religious will say God, but in scientific terms there are only two possibilities: an infinite cosmic consciousness or we humans. In fact, both possibilities merge into one. The universe exists only as humans experience it, and our ability to experience it comes from infinite or cosmic consciousness. This merging occurs because the source of consciousness, cosmic and individual, is within us.

I've taken this time to venture into some very abstract concepts,

Rationality has built a world of its own. In advanced societies the average person cannot enter this world without extensive specialized training. But there is a human side to all of this. A worldview that leaves the average person out in the cold makes for a very lonely existence. The newest iPhone isn't a hug. Einstein humanized the isolation of modern life in one of his most powerful quotes. A human being, Einstein said,

> experiences himself, his thoughts and feelings as something separated from the rest, a kind of optical delusion of his consciousness. This delusion is a kind of prison for us, restricting us to our personal desires and to affection for a few persons nearest to us. Our task must be to free ourselves from this prison . . .

That's one scientist's cry for something to unify us, bring comfort, and end the delusion of lonely isolation. In this book I am only expanding on the same desire. Virtual reality stands for every mind-made limitation that needs to be undone. As long as we believe in virtual reality, meta-reality lies beyond our grasp. Everything must be done in the name of freedom, which rescues reality and ourselves at the same time.

# 5

## MIND, BODY, BRAIN, AND UNIVERSE ARE MODIFIED CONSCIOUSNESS

The title of this chapter encompasses a new creation story. It begins by positing that a new story is needed. The old story has sunk in deeply. Schoolchildren have heard of the big bang, which was conceived of in their grandparents' day—the British astronomer Fred Hoyle coined the term in 1949, while the concept of an expanding universe that began with a bang goes back to the Belgian astronomer Georges Lemaître in 1931. Strange as it seems, the whole quantum revolution basically had no creation story. Einstein and the other famous names in his generation accepted that the universe is, was, and always had been.

Behind the workings of genius devoted to the radical notions of general relativity and the quantum, the "steady state" universe, as it was known, hadn't advanced since the days of the ancient Greeks. The Book of Genesis, if viewed a certain way, had one up on modern physics—Genesis and the world's other creation myths were based on the belief that everything in the cosmos had to have a beginning. The big bang allowed physics to fall into line with that commonsense idea, and the current estimate of a universe 13.8 billion years old could be verified, in place of biblical time (which Sir Isaac Newton, a devout fundamentalist Christian, spent years trying to calculate, so that he would establish a scientific date for the Garden of Eden).

A universe that began with a bang satisfied the commonsense mind, but it's a wobbly story once you ask where the big bang came from. An enormous problem blocks the way to finding an answer: logic breaks down when you ask what preceded time and space, since nothing can come "before" time or "outside" space. The inability to talk about a domain outside time and space, which nonetheless made the decision to create time and space, forced physics to retreat into arcane mathematics that doesn't depend on being able to form pictures or logical statements in the mind. As sci-fi fans know, if you go back in time and kill your grandfather before he got married, suddenly logic collapses in that instance, too. You cannot be alive (as a result of your grandfather having children) and dead (as a result of your grandfather not having children) at the same time.

A new creation story must get around this enormous, messy collapse of logic. Either it must describe a rock-solid ground zero when creation began, as Genesis does, or it must substitute something that doesn't need a beginning. The second path is the one I've been taking by grounding the universe in consciousness. Of all the candidates for an eternal something that has no beginning or end, consciousness is the most viable. In the new creation story, there are no gods or God, no wind-up universe running like a finely tuned watch, and no birth pangs in the big bang. Without them, there isn't even a story in the usual sense. There is only consciousness doing its thing while *Homo sapiens* generates creation stories that attempt to grasp what is ungraspable, to conceive of what is inconceivable. If God created the world, who created God? If the big bang gave rise to time and space, this implies a state of no-time and no-space, which is ungraspable. You can't logically ask what happened before time began, since "before" only has meaning within the framework of time. Creation wouldn't exist if we had to explain it with common sense first.

But there is a way to get there. Two people can disagree about everything in the world. One can dislike what the other likes. One can believe in God and the other be a confirmed atheist. But even if person A dedicated his life to contradicting everything person B said, even if he could get inside person B's mind and contradict every thought he had, on one thing the two antagonists would have to agree: They are conscious beings. If intelligent life from another solar system made contact with Earth, no matter what the creatures looked like physically—mutant humans, walking squid, amoeba-like blobs—they would have to be conscious, too.

The deepest mystery about alien intelligence won't be the technology it has developed. No doubt an interstellar spaceship or teleportation device or time machine will be mind-bending in scientific terms. There's a good chance that Earthlings could learn amazing things from alien technology. But the deepest mystery will remain unsolved. We will never know what it is like to be them, because their species of consciousness will be impenetrable. Little green aliens might feel physical pain from touching water; they might inform us that photons are delicious or that gravity is having a bad day.

Even those strange statements assume a lot. Little green aliens might not possess any of our five senses. As long as they are conscious beings, the reality they inhabit can take any form, because consciousness takes any form already. The most common butterfly on Earth is the painted lady, whose wings are brightly patched with orange and whose genealogy is delightfully poetic: the painted lady belongs to the group of *Cynthia* butterflies, a subgenus of *Vanessa* butterflies, which are part of the family *Nymphalidae*. If these names evoke a kind of ethereal mythology, populated by fairy folk, the painted lady's sense organs are as mind-bending as any alien's.

Painted ladies can taste the leaf they are standing on with the tips of their legs. They smell the air through their antennae and view the world through eyes that possess thirty thousand lenses. They hear with their wings. Looking from the outside, we'd say that evolution has created a reality for the painted lady that we barely comprehend. If you want to believe in aliens among us, believe in butterflies. As with visiting aliens, we'll never *experience* their consciousness.

But life is impossible without consciousness, which is another thing in favor of the new creation story. Fritjof Capra is an Austrian-born American physicist who gained fame with his 1975 book, *The Tao of Physics*. It was a landmark in the "new physics" because Capra connected science to ancient wisdom traditions like Taoism. Evidence from the subjective world, long considered useless in science, suddenly became relevant.

Capra has connected life and mind far beyond what biologists currently understand. A biologist would call the painted lady butterfly mindless. To the contrary, Capra asserts, "The interactions of a living organism—plant, animal or human—with its environment are cognitive interactions. Thus life and cognition are inseparably connected." In everyday language, there cannot be life without mind. The painted lady's sense organs are strange from the human standpoint, but it is a conscious creature. There is no other alternative explanation. Capra goes on to say as much: "The process of cognition—or, if you wish, of mind—is immanent in matter at all levels of life. For the first time, we have a scientific theory that unifies mind, matter, and life."

This sounds exactly like a metahuman declaration. It is human to believe that only *Homo sapiens* is conscious, with a small proviso added to include higher primates like gorillas and chimpanzees (maybe). It is metahuman to be aware that consciousness is universal. If this insight doesn't cause mental overload, that means its implications haven't fully sunk in. Consciousness can be described in three words: everything, always, and everywhere. On their own, each word is mentally unmanageable. If children took seriously that Santa Claus is everywhere, watching each boy and girl in the world to see if they were naughty or nice, the idea would be as thorny as a medieval theologian trying to grasp God keeping tabs on every sinner in the world, down to each person's secret thoughts.

There are reassuring ways to get around the dilemma and any possible anxiety. For every idea a human being has ever had, there are infinite ideas yet to come. When I realize this, I feel optimistic—vistas of unlimited creativity open up in my mind's eye. But I know I am only using an image; I am not looking at the reality of consciousness directly. There's a haunting passage in T. S. Eliot's poem "Burnt Norton" in which a bird summons us to run through a garden gate into "our first world," only to have the bird warn us back: "Go, go, go, said the bird: human kind / Cannot bear very much reality."

It's a sweeping and gloomy notion that humans cannot endure a full blast of reality, as if some kind of overload would blow our fuses. I don't agree. Our first world isn't the Garden of Eden, although that's the metaphor Eliot is using. Our first world is pure consciousness, which we can call "pure" because it doesn't contain anything. The open space between thoughts doesn't contain anything, either, except the potential for the next thought and the next and the next. Apparently, a kind of nothing, the gap between thoughts in reality is the "stuff" that a thought is made of. The mind is the play of consciousness once it starts making things.

I'd consider alien life forms to be superior beings only if they comprehended "everything, always, and everywhere." Short of that, we cannot enter metareality ourselves. The play of consciousness is behind all of creation. I think of this as "Genesis now," the constant appearance of newness in every dimension—mind, body, brain, and universe—that gives us the experience of being human.

"Everything, always, and everywhere" is the real story of creation. It alone tells us who we are and why we are here. Beyond the narratives provided by any storyteller, whether they are the rabbis of 1000 BCE compiling the Hebrew Bible, the mythologizers of Homer's Greece, or modern theorists of a gazillion possible universes, *something* is making creation happen. This *something* is here and now. It has no story in the conventional sense. It is content, like a choreographer standing offstage, to invisibly invent what the dancing universe is doing.

# The Ultimate Fail-Safe

Consciousness isn't playing *with* the universe. Consciousness *is* the universe. It plays at turning into a helium atom or a galaxy. It plays at turning into a brain cell or a heartbeat. Shape-shifting never ceases. When you feel happy or sad, you are experiencing two contrasting modes of consciousness. But since you are also consciousness, and so is your brain, there can be only one conclusion. Creation is consciousness experiencing itself. Like making all the gold jewelry in the world, the forms in creation change, but the gold—the essential "stuff" of creation—remains the same. In humanity's search for Truth with a capital T, this truth alone qualifies.

Is Truth with a capital T exciting, fascinating, or merely mind-boggling? When people get wind of consciousness, their worldview isn't shattered. They don't sit down in amazement as Truth washes over them. In fact, the average person shows complete indifference (in my experience) and dismisses such thoughts from his mind. The opportunity to access metareality doesn't strike people as anything urgent or necessary. I've reflected on why this is, and it comes down to a series of fail-safe mechanisms. A fail-safe keeps disasters from shutting a system down or making the unthinkable happen, as in the accidental launch of ballistic missiles armed with nuclear warheads. There are strong fail-safes to prevent such a catastrophe.

In the case of the human mind, the fail-safe protects virtual reality from being dismantled. Strangely, the mind must be protected from its own infinite potential. Imagine that you are an art student, and on the first day of class the instructor outfits your skull with electrodes that will fill your brain with every painting that has ever existed or ever will. You reluctantly comply, and suddenly the full reality of art pours into your cranium, from the earliest cave paintings on, covering thousands of years. Such an experience would be unbearable, in all likelihood, but it would also be useless. You can't teach budding artists by showing them every piece of art. It's like asking for a glass of water and being made to swallow the Great Lakes. In other words, the slow unfolding of the mind in time is a fail-safe. In linear sequence, event A is followed by B, which is followed by C.

This linear process is not real, however; it is a mechanism installed in human consciousness. One can argue as to whether our ancient ancestors installed it early in prehistory or it was installed by the force of evolution. But there's no doubt that we consider this mental fail-safe necessary for survival. One piece of evidence is how hard it was to convince ordinary people that time is relative. Even though the concept dates back to Einstein's theory of special relativity in 1905, the notion of time deviating from a straight line in a fixed way seemed unfathomable. As late as twenty or thirty years later, it seemed like a kind of special magic. The British philosopher Bertrand Russell, who wrote a small book titled *The ABC of Relativity* in 1925, announced publicly, with typical lack of modesty, that he was one of only three people in the world who understood what relativity was.

Einstein's revolutionary theory didn't alter everyday life; relativity could be set aside in its own exotic cage. (But there are practical applications. For example, the GPS satellites that orbit the Earth have to take relativistic effects into account. If they didn't, the GPS in your car would be off by a fraction of a second and thus give you an inaccurate location for where you are.)

Once you accept the human construct of time unfolding in a straight line, it naturally follows that there is cause and effect. The big bang led to the creation of planet Earth after around 10 billion years, which in turn led to the creation of DNA, then human beings, then civilization, then New York City, then the birth of a new baby in a New York hospital at some hour this morning. The reverse cannot be true—the birth of a baby in New York City cannot lead to the big bang. That would defy cause and effect.

This fail-safe is so convincing that we cannot easily accept its artificiality. But relativity wasn't the last or the only glitch. Modern physics has devised mathematical models of the universe in which everything happens simultaneously at the subtlest level of creation, where time dissolves into the timeless and the entire cosmos is a single subatomic particle. But such models are exotic, even in quantum physics, which is quite exotic to begin with. Nobody expects such notions to rise to the surface of everyday life. Without our fail-safes, we become very anxious.

We are all quite comfortable with the illusions we have protected with fail-safe mechanisms. We see the universe as a theater of space and time in which objects and people bounce around. Our bodies are objects. Our minds are the product of an amazing thinking machine called the brain, another object. Marvin Minsky of MIT, one of the fathers of artificial intelligence, defined human beings as "nothing but *meat machines* that carry a computer in their head"—a merciless expression of what most people assume to be true.

But there are also personal fail-safes that each of us constructed on our own. These personal fail-safes have been dubbed "the things that must be so." Some people must be in charge; others must win; still others must never start an argument. Is there a rational reason behind "what must be so"? No—these psychological mechanisms exist as forms of self-defense. They give us a sense of security, and even a seemingly small incident can trigger a major alarm. In *Love's Executioner*, a 1989 book by Stanford psychiatry professor Irvin Yalom, the author describes a middle-aged woman he calls Elva who had been the victim of a purse snatching outside a restaurant. The crime was random, and while she lost $300, there was nothing else in the purse that couldn't be replaced. Most people would feel the shock of such an assault and then move on. But Elva couldn't: "Along with her purse and her three hundred dollars, an illusion was snatched away from Elva—the illusion of specialness."

Having lived a life of relative privilege, Elva assumed that she was immune from this kind of disruption—she kept repeating, "I never thought it would happen to me." But it had, and the loss of illusion exacted a terrible price:

> The robbery changed everything. Gone was the coziness, the softness in her life; gone was the safety. Her home had always beckoned her with its cushions, gardens, comforters, and deep carpets. Now she saw locks, doors, burglar alarms, and telephones.

Elva was constantly anxious, and time didn't make her feel that her life was returning to normal. As Yalom explains it, "Her world view was fractured . . . She had lost her belief in benevolence, in her personal invulnerability. She felt stripped, ordinary, unprotected."

Being the victim of a crime shatters personal boundaries, because the crime is a personal invasion. Here, Elva couldn't repair her boundaries. She went into an existential tailspin, haunted by the ultimate threat: death. It wasn't a fear she had ever confronted, and yet death had touched her intimately when her husband died. It was a loss she had never really come to grips with, and it became the key issue that Yalom worked with her on in therapy.

Elva turned out to be an extraordinary patient, and in time "she moved from a position of forsakenness to one of trust," a change Yalom considers not just transformative but redemptive. Elva, who once thought she had created the perfect story, learned how to be *aware* of death without fearing it.

It's futile to believe that you can create a perfect story, which is most people's notion of a perfect life. Stories are always intruded upon by unwelcome events. And stories become tangled webs even when they seem to be working out well, because unconscious forces—anxiety, depression, anger, jealousy, loneliness—can erupt at any time. Although Yalom doesn't use the term *fail-safe*, one can see that Elva's sense of being safe and protected was a psychological fail-safe. It had served her well for a long time, until she discovered how flimsy and unreliable it actually was.

All of us have created our own version of personal fail-safes. As with most mental constructs, we often don't realize what we have done, and sometimes the mechanism seems to be so unconscious that it is totally out of our control. Phobias are a good example. People who are afraid of spiders will not physically be able to touch even a small, harmless one. If they reach out and attempt to touch one, fear will arise, growing more extreme as the hand gets closer. The next stage is violent trembling, a cold sweat, signs of panic, and the feeling of being about to faint.

Phobias offer perverse evidence about the creativity of consciousness, because anxiety can focus on anything. At the website The Phobia List, you can find a catalog in alphabetical order of the phobias specified in the psychiatric literature. A total number isn't given—new references are constantly being added—but under the letter A alone there are sixty-five entries, including *aulophobia* (fear of flutes). Not every letter is this full; under G, for example, there are only nineteen phobias listed, including seemingly silly ones, like *geniophobia* (fear of chins) and quite serious ones like *genophobia*, a fear of sex. No consensus exists for why phobias appear, although psychiatrically they are categorized as a form of anxiety disorder, the common element being excessive fear. Besides specific phobias, there are social phobias that involve situations where the person is afraid of what other people think, and a whole class of agoraphobias, where people grow panicky because they feel they are in a situation they cannot escape.

People suffering from phobias did not necessarily have a bad experience with the object they fear—they generally haven't—but they will go to great lengths to avoid the object once the phobia has emerged. Phobias seem strange to anyone not suffering from one, in part because avoiding the feared object or situation can lead to odd behavior, as with agoraphobes, whose fear of open spaces prevents them from leaving the house for years at a time. It is also strange that fear—which puts mind and body on high alert and is an effective protection when actual danger arises—turns counterproductive in phobias. Being on high alert because you hear a flute or see a chin has no survival value whatsoever.

Yet the normal fail-safes that we hardly notice have a breaking point, and when it is reached, severe dislocation occurs. Going into shock, for example, renders the victim of an auto accident, fire, or crime utterly helpless. The dazed look and inability to make decisions that are symptoms of shock could be seen as the opposite of fight-or-flight. In fight-or-flight, a flow of stress hormones is released, and under their influence, the lower brain is totally in charge—a soldier running from battle in a panic cannot consciously will himself not to flee until the burst of adrenaline subsides and the higher brain can once more make conscious decisions. (We can see the same response in a trivial situation, like being shown an illusion by a street magician. When he guesses the right card or pulls a quarter from behind someone's ear, it is quite common for the person to jump back or turn away in alarm—the lower brain is reacting to the trick as if it were a threat.) Shock is involuntary and would seem to have no self-protection value.

Gathered together, these various fail-safes are essential to the self-model, the view of reality that tells you what is real *for you*. No doubt personal reality has its quirks, and no two people inhabit the same self-model. But, collectively, we share virtual reality, with its universally accepted features of time, space, matter, and energy. These features are so ingrained that we feel "at home" as long as space, time, matter, and energy don't go haywire. But knowing that these are mental constructs isn't about reality going haywire. It's about acquiring self-awareness, which alone can tell us what is really going on.

# Buying In, Opting Out

No single individual created the human experience of space, time, matter, and energy—they were created in our collective consciousness, which goes back as far as the emergence of *Homo sapiens* and certainly before that. We cannot follow a trail of footprints that will retrace the steps that gave humans self-awareness, which is our species's unique feature. Our genes contain the evidence of every life form, including bacteria, that contributed to our physical makeup, but there are no physical traces to tell us how our ancestors experienced their lives.

Our species's most important inheritance is invisible. We connect with this legacy in early childhood, absorbing the whole virtual-reality setup. Once a young child learns that time exists, the rules of time soon follow, and then there is no turning back. To give an analogy, once you learn to read at age six or seven, you cannot return to the state of illiteracy. Letters on a page cannot turn back into meaningless black marks. Likewise, once you and your brain have adapted to the rules of time, it would seem impossible to live as if time didn't exist. A day, an hour, a minute, a second—this dissection of life into bits of time is what T. S. Eliot lamented in another poem, "The Love Song of J. Alfred Prufrock." Time has become Prufrock's psychological enemy.

> *For I have known them all already, known them all:*
> *Have known the evenings, mornings, afternoons,*
> *I have measured out my life with coffee spoons.*

Making a habit of dividing life into units of time doesn't mean that human time is more real than another version. We have no idea of how other species of consciousness experience time and space. Does the lumbering Galapagos tortoise feel it is moving slowly or the jackrabbit that it is bounding fast over a western prairie? We might speculate that animals live in the present moment, reacting to the instincts that tell them that now is the time to eat or sleep or hunt. But "present moment" doesn't exist for a creature that has no concept of time.

Nature gives us biorhythms that are imprinted in our genes, like the circadian (daily) rhythm for waking and sleeping. But that doesn't answer the riddle of how time exists in the first place. There is timing that is precisely tuned over thousands of miles of ocean, in a way no one can explain. For example, a small sandpiper known as the red knot (*Calidris canutus rufa*) migrates every year from Tierra del Fuego, at the southern tip of South America, to its breeding grounds in the Canadian Arctic. In a lifetime the red knot flies an estimated 240,000 miles, longer than the distance from the Earth to the Moon.

Why birds cross from one hemisphere to another to mate seems inexplicable, but the red knot embodies a specific mystery all its own. Needing food along its immense journey, the bird stops off in May on the beaches of Delaware Bay, timing its arrival with a primordial event. Between the full moons of May and June, hordes of horseshoe crabs emerge from ocean shallows to lay their eggs. A female lays between 60,000 and 120,000 eggs in one spawning. Dating back 450 million years, or 200 million years before dinosaurs, horseshoe crabs look like hard, rounded turtle shells with spiky tails (they are not actually crabs but are closely related to spiders and scorpions).

Within two weeks the eggs will hatch, so the opening for the red knots to feast on them is very brief. Yet every year, the stirring to migrate is felt 9,300 miles away near the Antarctic. The full moon in May can fall on any of the month's thirty-one days, and the red knot must start its journey from Tierra del Fuego in February. How did Nature synchronize the Moon, the mating cycle of a living fossil in the sea, and a tiny bird that has the longest migratory path on earth? In the red knot's case, the timing is somehow rooted in its DNA, and it comes with sweeping physiological changes. Prior to migrating, the bird's wing muscles expand while its leg muscles shrink. Because horseshoe crab eggs are soft and easily digested, the red knot's stomach shrinks prior to taking off. Its crop, which has grit in it to grind up the hard food that the bird eats in winter, also shrinks.

In other words, the red knot's DNA knows in advance every detail of the future. Since their Arctic breeding ground is bare, exposed tundra with no food source, the birds double their weight by adding fat during the Delaware feeding frenzy on horseshoe crab eggs, which lasts from ten to fourteen days. If any element in this synchronized cycle falters, survival is threatened. (Unfortunately, this has happened. Red knots are endangered for various reasons, a major one being the drastic decline in horseshoe crabs as coastal waters are damaged.)

Human beings are not tied to instinct when it comes to time, although our DNA somehow knows how to time puberty and menstrual cycles, for example. Cells are programmed to die at a certain time, a process known as *apoptosis*. A typical cell can only divide around fifty times before dying (the so-called Hayflick limit), and the process can be scientifically measured in the laboratory. But at bottom it is totally mysterious. Cells are chemical factories enclosed in a soft, permeable membrane. Chemical reactions happen instantly when two molecules meet; there is no hesitation, delay, postponement, or going back. How, then, does a collection of chemical reactions, each tied to the present instant, acquire the ability to time future events? The question is so basic that almost no one asks it.

Imagine a game of pool, with the balls representing atoms and molecules set to collide inside a heart or liver cell. After you break the pack with your first shot, the rest is mechanical. When pool balls collide, they must bounce off each other instantly, and where they travel next is determined by Newton's laws of motion. As one of Newton's laws dictates, once a ball is hit or collides with another ball, it must travel in a straight line until something stops it. Nothing abnormal seems to happen as the game unfolds. Yet when you return to the pool room the next day at noon, the balls have already arranged themselves in a pack, ready for the first hit. Such behavior would be remarkable enough in pool balls, but your DNA anticipates countless events, controlling the timed release of hormones, for example, anticipating the need to refresh the chemical triggers that fire in a brain cell with every thought, and organizing hundreds of synchronized biorhythms with absolute precision unless we interfere (by staying up all night, for example, or taking hormone supplements).

We cannot enter into the experience of time that other species have. But since time is so variable and malleable, we can say that it is constructed differently by the DNA of each living thing. Therefore, it takes no leap of imagination to say that time is a construct to begin with. Since DNA consists of atoms that interact instantly, something outside the atom must be doing all this precision timing and synchronization. The only viable candidate is consciousness. After all, you cannot manipulate time unless you know that time exists.

## Living in a Creative Universe

If time is a construct, the same is true for the other things we consider essential to the universe. Our experience of matter and energy is quite species-specific. Does a termite steadily gnawing its way through a house's wooden frame experience the wood as hard? Does a mole confined underground for its entire lifetime in a tunnel barely wide enough to crawl through experience space as tight?

For our species of consciousness, time, space, matter, and energy are malleable experiences, tied to our creativity. That is, they can expand and contract in various ways, as Einstein explained with a quip: "Put your hand on a hot stove for a minute, and it seems like an hour. Sit with a pretty girl for an hour, and it seems like a minute. That's relativity." But he was actually sidestepping the core issue of time: Is it relative just because humans say so? Sitting in a dentist's chair dislocates time for people who fear the experience; they cannot say how much time has passed, except that every minute was unpleasant. This doesn't alter the hands on the clock in the dentist's office, so which version, the personal experience or the mechanical device, is real?

It will seem bizarre, but human experience is what makes time real, not clocks. To get at how we create the experience of time—along with space, matter, and energy—you must opt out of accepting that these are fixed things. The current theory of the universe helps us out here, because after Einstein proved that matter can turn into energy ($E = mc^2$), the door was opened for other transformations. A later physicist, the American Richard Feynman, could even show mathematically how an electron's position can be expressed as moving backward in time. This is like being asked where you live, and you answer, "I live at 63 Maple Street, although sometimes I live in last February. Take your pick." Our stable notion of cause and effect has been dismantled by experiments in "reverse causation," where an event in the future affects what is happening now.

The ultimate transformation, however, occurs when the field of virtual particles morphs into physical particles. This is commonly known as "something from nothing." Virtual particles are invisible and have no location in time and space, but they are totally necessary for the physical universe. It's like reversing ghost stories, where the ghost comes first and the living person second. If you look at your hand and begin to reduce it to finer and finer levels of physicality, it quickly becomes a shivering web of molecules. These molecules are not as solid as your hand, and at the next level down, you arrive at a congregation of atoms that are barely physical, being over 99.9999 percent empty space. This is the last level at which physical existence clings by a thread. At the level of subatomic particles, there is a winking in and out of existence as quarks, gluons, and other exotica of the quantum domain shift from being virtual to being intact in our physical universe. Going from nothing to something is happening in your hand and in every other physical object all the time.

So the pivotal issue isn't that solid physicality is an illusion. No one can dispute this—we couldn't exist without buying into the psychological security blanket that the world won't vanish tomorrow in a puff of subatomic mist. The pivotal issue is whether consciousness, and particularly human consciousness, is the creative force behind "something from nothing."

Being human unfolds in time and space. Your birth certificate attests to the date of your birth and the town you were born in. No one's birth certificate says, Date of Birth: Eternity; Place of Birth: Everywhere. The commonsense assumption is that time and space simply exist "out there" as part of the natural world. To evolve to metahuman, however, we need to stop thinking of creation in fixed terms. Reality must be rebuilt to account for the role consciousness plays. There are only two levels of reality. One level is unbounded pure consciousness, which is a field of potential. The other level is consciousness in its excited state (to borrow a phrase from particle physics), which we call the universe.

All excited states vibrate with energy. Matter is vibrating with physical energy, the mind with mental energy. The body is an excited state; so is the brain, being part of the body, and so is the mind as thinking goes on. Even when a rock looks like a rock and a neuron like a neuron, which makes them seem totally dissimilar, both are excited modes of consciousness.

This is practical knowledge. When you know, for example, that time is just an excited state of consciousness, you see why time is so malleable. It has no choice but to be as fluid and flexible as the mind. We manipulate time to conform to our human needs. Some time frames are long, like the life of the universe, while some are short, like the milliseconds it takes for a signal to jump from one nerve cell to the next. These time frames can be extremely stable: hydrogen atoms are likely to exist until the death of the universe, while a thought is stable only for its duration, which is transient and evanescent. The fact that mental activity flies by so fast can lead to a mistake. We might suppose that consciousness is on a time schedule—it isn't. Consciousness can be fast or slow, large or small, random or predictable, and so on.

Being the source of "something from nothing," consciousness isn't bound by its own creation, just as a person with a vocabulary of thirty thousand words isn't tied down to thirty thousand thoughts. Combining and re-forming all the time, consciousness dictates to itself whether to be as tiny as a quark or as immense as the universe. This fact helps dispel one of the most obstinate objections to consciousness-based reality. We don't see our minds create trees, mountains, planets, and stars. The scale is wrong, skeptics will say. When you are afraid and your heart starts to pound, a mental event—your fear—sets chemicals in motion in your body. The scale is molecular, meaning very small. Moving a mountain with your mind, however, cannot happen, because it is too large.

This objection is invalid, however, because consciousness doesn't respect limits of scale. Imagine that you wake up from a nightmare, and you tell a friend that a hundred men with guns were chasing you down the street. What if your friend replied, "I'd buy it if one man were chasing you down the street, but a hundred is too many to be believable." Such a comment would show ignorance about how dreams work. Dreams are not bounded by big and small. A mouse chasing you down the street during a nightmare is the same as the Red Army invading your town. In a dream a blade of grass can tremble, followed by a planet exploding.

We accept these anomalies in dreams because we are used to waking up and returning to the physical world and its constraints. Consciousness is set up so that some things are free to move around just by thinking, like brain chemicals that move around in sync with our thoughts, while other things aren't movable by thinking. That's the setup in a human universe. We don't know the limits of the setup until we test them. One has to ponder if Jesus had a similar understanding in mind when he declared to his disciples, "Truly I say to you, whoever says to this mountain, 'Be taken up and cast into the sea,' and does not doubt in his heart, but

believes that what he says is going to happen, it will be granted him"
(Mark 11:23).

Whether you accept those words as gospel or metaphor, literal truth
or a vivid teaching example, the notion of moving mountains with your
mind sounds supernatural, therefore irrelevant to everyday life. The
problem isn't that supernatural = impossible. The problem is that, so far,
science cannot explain what we call "natural." Leaving consciousness out
of standard scientific explanations dooms them to failure. The mind cannot be explained by juggling brain chemicals around, and when science
declares that there is no other explanation, we are no longer in the realm
of credibility—we are in the realm of "things that must be so."

Once virtual reality has been dismantled, however, there is no going
back. You cannot see through an illusion and believe in it at the same
time. A magician can't sit in the audience and be fooled by his own tricks.
However, that's exactly what we do. We place our faith in the physical
world while knowing full well that it is illusory.

Instead of passively accepting the commonsense world, metareality
gives us an alternative—seeing everything in the universe as shifting
modes of consciousness. A tree, for example, is tailor-made to fit our
human response to it. Any quality of a tree can be taken out of its assigned slot and reassigned to fit a different, nonhuman framework. The
tree's color doesn't exist for someone with complete color-blindness. The
solidity of a tree doesn't exist for a neutrino, a subatomic particle that can
whiz through the Earth as if passing through outer space. The weight of
the tree doesn't exist if you transfer it to the International Space Station.
The life span of a tree vanishes when viewed from the life span of the
protons in the nucleus of every molecule, which take billions of years to
decay.

Everything acquires its "realness" from the mode of consciousness
applied to it. In sleep the entire physical world disappears and no longer
exists *for you*. It still exists collectively, held in place by the rules of virtual reality. But in sleep you opt out of virtual reality by experiencing a

different world, which isn't the blank unconsciousness most people assume sleep to be. It is possible to experience deep sleep as pure awareness without excitations—indeed, that's how centuries of swamis, yogis, and other versions of metahuman experience it. The Buddhist concept of Nirvana is closer to deep sleep than to the ordinary waking state, because Nirvana reconnects a person with pure awareness.

Dreaming is a mode of consciousness in which mental excitations are subtle. This is yet another world, one where the rules of virtual reality don't apply. In dreams, objects magically defy the ordinary laws of physics—a locomotive can fly, the Empire State Building can vanish in a puff of smoke. There is no reason to demote those two modes of consciousness, sleeping and dreaming, to an inferior position compared with being awake. If we can look at the world as a lucid dream, why consider the dreams you experience at night as any less real or unreal?

People blindly follow the assumption that hard physical things are more real than subtle things like thoughts, imagination, and dreams. But your subtle impulses determine how your personal reality operates. I've given one example—phobias—where a person who can lift hundred-pound weights at the gym may be unable to lift a spider in his hand; fear freezes up his muscles. It doesn't matter how light the feared object is. But we also need positive examples of how a person's subtle intentions can alter personal reality.

Think back to our earlier example of using VR apparatus to simulate the experience of standing on a skyscraper girder high in the air (page 36). A person standing safely on the ground, thanks to the VR illusion, feels giddy and weak, and experiences a powerful threat of falling. This indicates that our sense of balance can be consciously manipulated in various ways. Tightrope walkers have separated their sense of balance from any feeling of threat or danger. To someone standing on the ground looking up, the danger looks real enough (your heart can pound simply from watching a risky circus act without experiencing it firsthand).

In the accepted model of evolution, all the traits we inherited have survival value. Our ancestors needed a sense of balance for obvious reasons when hunting and fighting other creatures in the wild. But we go beyond survival all the time and toy with our evolutionary inheritance simply because we want to. Tightrope walking has no survival value, and since it poses considerable danger unless you are highly trained, it has negative survival value.

Babies are afraid of falling from a very early age, and they can't learn to walk without testing the precarious state between falling down and staying on their feet. Clearly fear of falling loses out in the end. Tightrope walkers go a step further by consciously overriding evolution when they disconnect fear of falling from sense of balance. This ability to override evolution is actually a higher evolutionary trait.

The power of consciousness allows daredevils to hang glide or free-climb vertical rock faces using any rationale that comes to mind—thrill seeking, a drive to accomplish the impossible, competitive rivalry, or simply no reason at all—oblivious to their life-or-death situation. Such freedom is absent in creatures hemmed in by physical evolution. Only *Homo sapiens* turns extreme risk-taking into fun. We choose our own motives, and personal reality moves with our intentions. It would be horrifying to find yourself in a dark alley facing an assailant armed with a knife, but someone with appendicitis, blocked coronary arteries, or a tumor willingly undergoes the controlled violence presented by knife-wielding surgeons. Once consciousness interprets a situation in a specific way, the reality of the situation fits whatever consciousness has decided.

The virtual reality that surrounds us permanently is a construct that fits so many motives, intentions, decisions, and interpretations that we've lost track of them all.

. . . . . . . . . . .

This chapter has dived into deep water, and some issues get complex. But the purpose of the chapter is simple—to close the gap between illusion and reality. No one's body is going to vaporize. The commonsense world will be there to greet you when you wake up tomorrow morning. Yet everything is entangled in illusion, making the commonsense world unstable at its core. By testing the very limits of time, space, matter, and energy, we are testing our own creative power. Moving a mountain with your mind is nothing next to moving the whole world, which is our ultimate goal.

# 6

## EXISTENCE AND CONSCIOUSNESS ARE THE SAME

I have strong memories, growing up in India, of attending schools taught by Catholic missionaries. The Old Testament wasn't light reading—I came away remembering sin, wars, laws, lamentations, and plagues. But there's a verse very early on, after the seven days of creation, in which God "was walking in the Garden in the cool of the day" (Genesis 3:8). God was still on good terms with Adam and Eve, and, in my imagination, the moment was perfect for a casual conversation. None occurs. Adam and Eve aren't actually there at first, having hidden from God out of shame for their nakedness. But in my imaginary conversation, they would ask a crucial question: "Why did you do it? Why did you create the world?"

In reply, God would say, somewhat sheepishly, "There wasn't any reason. I just had to. I couldn't help myself."

Modern cosmology has come up with no better answer. The universe is self-creating. It exists just because it has to exist. Everything since the big bang has unfolded of its own accord, driven by natural forces. We almost didn't make it. Physicists have calculated that the balance of creation and destruction was extremely delicate in the beginning.

Destruction came within a hairbreadth of winning out, because all but the tiniest fraction of primal mass and energy collapsed within itself, returning to the vacuum state, as it is known. As little as one part in a billion escaped the jaws of destruction—a metaphor for gravity, the force that caused the collapse of everything else—but one billionth of creation was enough to enable trillions of stars and galaxies to be formed. (This didn't happen quickly, though. It took 800 million years for the oldest star to coalesce from interstellar dust.)

A self-creating universe is our version, with many bells and whistles attached, of the eighteenth-century notion of God as a cosmic clockmaker who wound up the early universe, set it going, and departed to let it operate mechanically. But self-creation is a sticky business. Who or what got the process going? The creator's creator wasn't a problem for the ancient Jewish rabbis—they accepted God on faith as always existing. Finding a similar nonfaith agent has proved all but impossible.

I'm arguing in this book that consciousness is the only viable self-creator, turning itself into mind, body, brain, and universe. That's a radical departure from physical explanations of the cosmos. As in the Book of Genesis, consciousness creates because it has to. It only needs to exist in order to start the ball rolling. (Where did existence come from? We don't have to worry about that, because if nonexistence existed, it wouldn't be nonexistence.)

Just because it exists, pure consciousness generates reality as we know it. The ancient Indian scriptures speak of myriad worlds spinning into creation like motes of dust dancing in a sunbeam. But in the same sunbeam we see myriad thoughts, feelings, sensations, and images, too, the entire content of the human mind. Faced with staggering overload, our minds have to generate a livable world to inhabit—a human world—and the virtual reality that we created was inevitable.

But simply reducing the teeming multiplicity of raw reality wasn't enough. The human world contains meaning. Where did it come from? It was always there, as a primary trait of metareality. Pure consciousness created a universe filled with meaning. Everything we can say about being human—our joy, love, vitality, intelligence, and infinite potential— needed no creator. Those qualities came with existence, right off the bat. Without wheels, a car isn't a car. Without humanity, the cosmos isn't a cosmos we could ever relate to, or even inhabit.

The universe was created for us because there is no other alternative. Creation cannot be whole unless we are whole. Once this secret is revealed, metahuman can free itself from self-doubt, confusion, and sorrow. They are the stepchildren of illusion.

# A Special Creation

Pure consciousness cannot be interviewed, like talking to God, even in an imaginary conversation. But the notion of self-creation has always been viable. We can accept without question that *Homo sapiens* created its own version of reality. We are still at it, with no signs of stopping. It makes no sense to claim that our remotest ancestors somehow learned to be creative. Creativity is intrinsic to us, like breathing.

Somehow we picked up a lot of old habits that consciousness had gotten used to. All life forms are wrapped up in self-creation, but they are content to let it roll over them. The ancestor of all mammals was a shrew-like creature named *Juramaia* that lived in tree ferns around freshwater lakes 160 million years ago. The discovery of this creature in China in 2011 pushed the ancestry of all "true" mammals back by 35 million years. (A true mammal species is one in which the fetus is nurtured by a placenta in the womb. This accounts for 95 percent of all mammals, the other portion being marsupials like possums and kangaroos that give birth to tiny babies that finish their maturation inside the mother's pouch.)

*Juramaia* didn't know that it would be anything but a shrew, and you'd think, looking at its five-inch skeleton, that only modern shrews would descend from it, not all true mammals. No trace of dogs, cats, bats, elephants, or whales is visible. It took a lot of inference to suggest that *Juramaia* had a placenta, since no soft tissues are preserved in fossils. Yet guesswork isn't needed when it comes to knowing that a placenta was the key. Once it appeared, a unique feature never seen before, a door was opened. What rushed through it wasn't *Juramaia*, which eventually became extinct, but self-creation. An unstoppable habit took a leap forward. From then on, true mammals could be small, big, or mammoth. They could swim, walk, crawl, burrow, or fly. Nothing was fixed or permanent—except a placenta. Nature took a creative leap without precedent, with barely a hint in all past life forms that this new kind of birth was possible.

This brings up a second misconception, that evolution is all about progress. Creativity doesn't need to progress; it is already complete. Every life form is a complete creative act. In the case of *Juramaia*, a tiny shrew-like creature wasn't better than what came before. Marsupials could have ruled the Earth—they became the only indigenous mammals in Australia. Much earlier, eggs evolved, and they kept on hatching, just as they did during the reign of the dinosaurs. Live birth wasn't better— today some sharks lay eggs while others give birth to live young. There is no straight-line progression from early life forms to us. Billion-year-old microbes, one-celled animals, and blue-green algae continue to exist because their adaptation to the Earth was perfect in its own way. Without the unstoppable urge to create, there was no reason for primitive life to leave its very secure comfort zone.

Self-creation is a strong clue that consciousness needs nothing outside itself to keep expanding endlessly. In time, *Homo sapiens* rushed through another door, the one that opened to self-awareness. We can do anything with our potential that we want to, but no matter how civilization changes, it is impossible to be human without being aware. The only question is the degree of awareness we choose to embrace. Metahuman is closer than we imagine, being just a higher degree of awareness.

Awareness and existence are uncreated. That's the simplest formulation of the truth. Being uncreated means they simply *are*. Without having to give a reason, needing no creation story, the primal setup for being human is that we are here. We exist because consciousness exists. Of all the secrets of metareality, this one takes the longest to sink in. At first glance, existence is a non-issue. Nobody in a college debating society like the Oxford Union will argue the case for nonexistence. There would be no debating societies if we didn't exist—the point seems silly, even childish. But if being here is enough to bring about everything in creation through a conscious process, *that* is big news.

Existence and consciousness cannot be separated. They don't just belong together, like heat and fire or water and wetness. They are the same thing. Descartes famously declared *Cogito ergo sum*, "I think, therefore I am." It's more accurate to say the reverse: I am, therefore I think.

Either way, however, seeking a cause-and-effect relationship is chop logic. To exist and to think are the same thing. One doesn't cause the other. Failing to see this fact has led to many misconceptions. For example, if you insist that the mind needs a cause, you are stuck giving it a creation story. If you are a modern secular person, you are soon trapped in the mistake of believing that the brain created the mind. To a developmental psychologist, there is no doubt that the brain creates the mind, and neuroscience concurs. As creation stories go, this one is bolstered by physiology, but childhood development also offers enough evidence to blow the story apart.

# The Expanding Brain

If we look at the last stages of the fetus's life in the womb, the brain is the main focus of development, the last organ to acquire its final configuration—but that won't happen yet. At the moment of birth, the infant brain is exploding with untapped growth potential. The biggest part of the human brain is the cerebrum, responsible for thinking and other higher functions, and evolution has given babies a huge cerebrum to begin with, one reason that a newborn's head is so large, at the limits of what can pass though the birth canal. In a normal fetus carried to term, the last trimester sees the brain roughly tripling in weight, from 3.5 ounces at the end of the second trimester to 10.6 ounces at birth.

In a sense, all babies are born prematurely, entering a "fourth trimester" the day they are born, because the brain keeps up its accelerated growth outside the womb. In the first three months of life, a newborn's brain grows by as much as 1 percent a day, expanding by 64 percent in the first ninety days, after which the growth rate shrinks to an average of 0.4 percent a day. During the rapid-growth phase, 60 percent of the energy consumed by a baby is used by the brain. The large-scale or macro view shows that a newborn already has all the brain cells needed for a lifetime—in fact, too many. A newborn's brain cells number around twice what an adult has, even though the newborn brain weighs half as much. As it grows to full size, which takes three years, the baby brain thins out its weaker connections (a process technically known as "synaptic pruning"). It's like sorting through an attic filled to the rafters with things you need and junk you can do without.

Here we begin to see the mystery of uniqueness, because synaptic pruning is different for each baby. The brain somehow disposes of what will not be needed *for that one individual*. A musical prodigy who needs phenomenal motor skills to play the piano at a virtuoso level may not need advanced mathematical ability or articulate language skills. A sculptor with a highly developed ability to envision objects in three-dimensional space may wind up without the relationship skills to find the right spouse. The combinations are endless, and, to match them, the brain's quadrillion synaptic connections are just enough. Yet clearly a brain cell has no foreknowledge that it will be needed or not needed, according to events far in the future. Synaptic pruning isn't random. It only makes sense that a higher perspective that is beyond time must be managing the process.

For example, a study by researchers at the University of Washington found that specific areas of the brain that control the physical aspects of speech (Broca's area and the cerebellum) are activated in seven-month-olds in advance of learning to speak. This is evidence of looking ahead to the future and throwing the necessary brain switches in advance. A seven-month-old baby doesn't know that she will one day speak, but *Homo sapiens* does, because we are a species of consciousness. Our species communicates through the spoken word, and just about every baby has inherited that ability.

This is different from inheriting the physical need to exchange baby teeth for adult teeth, which is linked to the growth of the jaw, or the need to go through puberty, which makes it possible to pass on our genes through sexual reproduction. Speech is a mental acquisition, a prime tool for knowing what someone else is thinking.

If understanding the brain would allow us to understand the mind, we'd be far ahead of the game. Yet the notion that "brain creates mind" has always been a fallacy. For a cell to know anything, it cannot get that knowledge from atoms and molecules. Atoms and molecules don't know that speech will be necessary. Only consciousness affords a valid explanation, because consciousness is the knowing element in every cell, every life form, and every person.

Because we live in the golden age of neuroscience, you would think that someday there would be a user's manual for the mind. But that day will never come. Our view is blocked by the mistake of equating mind and brain. Even the fact that the brain contains a quadrillion connections is meaningless in terms of mind, just as measuring every frequency of visible light would be meaningless to explain how Leonardo da Vinci painted the *Mona Lisa*. Life would be meaningless if the brain were in charge, because the brain itself is meaningless. The mystique that surrounds it, ascribing thoughts, feelings, and sensations to clumps of neurons, has no basis. If a neuroscientist went around saying, "That quarterback is out of his brain" or "I'd like to buy a house, but I just can't make up my brain," no one would deny that *mind* is the word intended.

It's very difficult, however, to pry neuroscientists from their belief that "brain creates mind." In the emergency room, an EEG reveals when the victim of a fatal car accident is brain-dead, at which point there is no mind left. Isn't it obvious that the mind lived there, beneath the hard shell of the skull? Not at all. Imagine that you had never seen a player piano and knew nothing about how one works. You walk into a music room, and a piano is playing "The Blue Danube Waltz" by itself. You can see the keys going up and down, the hammers striking the strings.

How would you know if there was an invisible pianist performing this magical act? There's no way to tell. On the other hand, it seems irrational to jump to the conclusion that the piano is playing itself. The instrument's raw materials—wood, steel, felt, and ivory—can be studied down to the atomic level. Nowhere will you find a talent for music. Yet in neuroscience and everyday life, we say that the brain, which is also a physical instrument, thinks, feels, sees, and does everything else that happens in the mind. We make this assumption without the slightest proof that the raw materials inside a cell—basically hydrogen, oxygen, carbon, and nitrogen—can love or hate Brussels sprouts, enjoy Viennese waltzes, fall in love, and so on. The fact that the brain is constantly buzzing with activity doesn't tell us that a mind is at work or why each mind is unique *at any given moment*.

Finding the source of the mind is easy once you put consciousness first. It's not hard to do this. We naturally rely on the mind because our existence is tied to it. We use our minds in lots of ways that do not light up on a brain scan and that require no discernible brain activity. Can you recognize your spouse's face in a crowd? Yes, instantly, but you don't go through a process that requires brain activity. Your brain doesn't flip through a mental Rolodex until you find a stored image of the right face and pick it out, the way the witness to a crime is asked to flip through mug shots at the police station. Recognition happens in the mind without burning calories in the brain, and unless the brain is burning calories, it's not at work.

In the same way, you choose words without thumbing through a dictionary stored in the brain. And once you learn the streets of a new town, you can negotiate them without consulting a map—on paper or in your brain. Learned things are simply there. The opposite is also true. If someone asks you the meaning of *hydrocephalus* or *ratatouille*, and you don't know the word, you don't need to comb through your vocabulary for missing words—you immediately know that you don't know. But a high-speed computer has to consult its stored memory before it says, "Does not compute."

Some years ago I met a math prodigy, now retired from being a professor, who had started publishing in leading mathematics journals at the age of twelve and subsequently attended Princeton at sixteen. He finished his coursework for a doctorate at Harvard while still an undergraduate. As remarkable as prodigies are, one comment in an interview with this math genius stuck in my mind. When asked how his thinking process was different from other people's, he said that he didn't think in order to solve a math problem—he posed the problem to himself, allowed it to incubate, and waited for the right answer to pop out.

He had developed such absolute confidence early on and still continues to rely on it. Should we consider genius exceptional, totally divorced from the norm? Not at all. To some degree everyone has experienced the power of intuition, which can be defined as the mind jumping to a conclusion without having to think through all the individual steps along the way. When people say intuitive things, like "I knew the man I was going to marry when I first set eyes on him" or "I knew I'd fly jets from the time I was five," they didn't arrive at this certainty by thinking it through. In all kinds of activities, the mind relies on the knowing element that is innate in consciousness. This fact must be grasped before we can venture into the metahuman possibility of using our full potential. The door is slammed shut if "brain creates mind," because what makes us human—love, compassion, creativity, insight, and imagination—is beyond the brain.

We can bolster the case by returning to the infant brain yet again. At birth a brain scan will indicate that a newborn's brain is highly active, but no one would claim that a newborn is thinking, not in the way adults put words together in their head. When an infant cries, it is inarticulately saying it is hungry, tired, afraid, needs changing, and so on. The mother picks one of the possibilities and acts accordingly once she pinpoints what her baby wants. We don't assume that the baby knows anything beyond its raw discomfort.

At some point, thinking begins, and an infant's life becomes mind-based. Thinking in words and forming ideas begin to emerge. But even if you were present at the exact instant a baby uttered her first word and even if you could map this moment as an event on a brain scan, the birth of language isn't in the brain. Somehow, purely in consciousness, a thought arises, connected to a meaning.

At first the use of words is like a parrot repeating what it hears. *Goo goo* means nothing. *Mama* is probably a parroted word coaxed out by the baby's mother. But then an invisible, silent *aha* occurs. *Mama* is the concept that attaches to a person, and only one person in the whole world. Then there's an acceleration of speech, and by their second birthday, infants enter what experts call a "meaning explosion." (This is also apparently the point at which chimpanzee DNA, although 98 percent identical to human DNA, stops short. *Homo sapiens* is unique in having billions of new neural connections developing in the first two years of life.) A typical two-year-old knows around a hundred words; at two and a half, vocabulary has grown to around three hundred words.

Meanwhile, a change occurs that is far more significant than acquiring a vocabulary: toddlers start saying things *they have never heard before*, forming new sentences on their own. No one taught us how to do this, and yet the most advanced computer is miles behind the average three- or four-year-old. A computer can be programmed to invent an infinite number of sentences never said before, but it doesn't know what it's doing. The process is totally mechanical—a laptop saying "I love you" doesn't really mean it.

Meaning can't explode out of the meaningless, just as sentences that make sense cannot emerge simply by tossing alphabet soup in the air. Consciousness unfolds as mind, body, and brain, each process governed by meaning and purpose. (A popular motivational slogan in recent years is "the purpose-driven life," which inspires people who feel confused, adrift, and without purpose. Yet we cannot help but be purpose-driven from birth—otherwise, we'd be mindless.) We may not be able to articulate the meaning of life, but we are certain that there is meaning, and we are designed to reach for more.

Consciousness is a meaning field. We know this despite ourselves. In one experiment, participants were told that they would be taking a hearing test, which required them to listen to recordings of sentences being said in a very low voice. Even with good hearing it was hard to make out exactly what was being uttered, but the participants were told to make the best guess they could.

What they didn't know was that each was listening to a sequence of nonsense syllables, a possibility that occurred to no one. No matter how low and unintelligible the voice, the participants heard—or guessed they heard—sentences that made sense. Yet even before language arose, animals made sense of their surroundings, learning to recognize edible food, potential threats, a desirable mate, and their own young. We might call this recognition instinctive, but it doesn't change the fact that meaning was inescapable. Even a one-celled amoeba engulfing a smaller one-celled creature is distinguishing food from nonfood. Nothing, not even the force of evolution, created meaning out of chaos—it is a given in the universe.

The mind field has the capacity to turn into a minefield, depending on what kind of mind a child develops. The people who hide from their own inner world, who dutifully shape their lives according to social norms, do it in order to avoid trouble. But hiding out from the mind's dark side doesn't defuse the threat. A tragic view of life hangs over us after the unthinkable carnage of the twentieth century, when an estimated 100 million people died as a result of war and genocide.

The cognitive explosion that anthropologists surmise occurred in prehistory, a leap that revealed the world of self-awareness, repeats itself with every baby's cognitive explosion as the baby learns to think and speak. The infant brain adapts to the worldview imposed on it. Infinite potential gets reduced in ways far too limiting and self-defeating. The British writer Aldous Huxley took a similar view decades ago. He wrote extensively against the use of the brain as a "reducing valve" that narrowed down the mind. Here's one of Huxley's most forceful statements on the matter:

> Each one of us is potentially Mind at Large. But in so far as we are animals, our business is at all costs to survive. To make biological survival possible, Mind at Large has to be funneled through the reducing valve of the brain and nervous system. What comes out at the other end is a measly trickle of the kind of consciousness which will help us to stay alive on the surface of this particular planet.

Using other words, Huxley is contrasting the unbounded field of consciousness with the tightly bound perspective the everyday mind takes. His central point, that the "reducing valve" of the brain was necessary for our survival, has gained modern support from various proponents of evolutionary psychology. They, too, view the narrowing of mind as a necessity for survival. But I've been arguing that consciousness never loses its infinite potential, no matter what is happening to us inside virtual reality. Placing the blame for our flawed existence on the brain points in the wrong direction. The brain is another mode of consciousness. Its potential isn't inherently limited. Let's go deeper into why this is so, because, for practical purposes, accessing metareality won't be possible unless the brain can break free from its narrowness and take us there.

# Mind at Large and Psychedelics

The first requirement is to open up the brain's reducing valve, which has been happening through a surprising avenue. A new wave of medical interest surrounds the potential value of psychedelic drugs, spurred by a thorough, sensible review of a once-taboo subject by Michael Pollan in his first-person account, *How to Change Your Mind*. For hallucinogens to resurface was a "Come out, come out, wherever you are" proposition. LSD, magic mushrooms, and mescaline had their day in the sixties and emerged from that time badly tarnished. Leaving aside various antidrug laws largely prompted by fear, a medical researcher who looked into psychedelics typically faced the risk of censure, perhaps career-ending censure. At the very least, such research wasn't taken very seriously and was quickly brushed aside.

The general view of psychedelics has been, up until recently, that they are potentially unsafe and medically useless. That is all changing. What has changed this conventional wisdom is deeper knowledge and better understanding of the brain. In particular, the area of the brain that seems to cause the mind-altering effect of LSD and company is the so-called default mode network (DMN), a collection of regions in the higher brain that organizes and regulates a wide range of cerebral activity. The DMN filters out the flood of information that bombards us every day, selecting and controlling our response to the world. It is the physiological location for the editing of reality we've been discussing in this book. A troubling implication of the DMN is that our brains evolved physically to become Huxley's reducing valve. LSD jolts the DMN temporarily, but once the trip is over, we go back to the brain's status quo.

Undoubtedly, the DMN serves a totally necessary function. Instead of feeling overwhelmed by a barrage of chaotic stimuli, a stable brain helps us approach life with a balance of judgment, experience, and self-interest. The DMN has been called the "me" network, because it functions in the brain the way the ego functions in psychology, tamping down irrational impulses and keeping them in check while organizing a balanced adult self.

The DMN doesn't develop until around age five, coinciding with the time when children become proud of "not being a baby." This entails many things—not throwing tantrums and crying over each little discomfort; showing more courage, independence, and self-possession; wanting to be useful by helping out; and defending their own tastes and inclinations. A great deal of inner self-regulation is required for such a complex shift in behavior, and the DMN handles the majority of that.

But this shift isn't altogether good and positive. From Huxley's perspective, Mind at Large is being reduced for the sake of crude animal survival. From a medical standpoint it is conjectured that the DMN, although totally necessary so that we aren't tripping all the time in a whirling blur of sights and sounds, has a downside. Over time its automatic (i.e., default) responses become ingrained and rigid. On the one hand, this may account for the stubborn narrow-mindedness associated with growing old, while on the other there may be a tie-in to disorders like anxiety, depression, and addiction. These become ingrained responses that won't be dislodged until the DMN stops controlling them so tightly.

One of the most searching accounts of personal transformation on psychedelics is *Trip*, the 2018 memoir of the American writer Tao Lin, born of Taiwanese parents in Florida in 1983. Lin had long been afflicted with despondency. As he puts it:

> Life still seemed bleak to me, as it had in evolving ways since I
> was thirteen or fourteen. I was chronically not fascinated by ex-

istence, which . . . did not feel wonderful or profound but tedious and uncomfortable and troubling.

Isolated, leading a hermit-like life whose solitude was aggravated by what Lin calls his addiction to the internet and his smartphone, he came to psychedelics by bingeing for thirty hours on the YouTube soliloquies of Terence McKenna, a passionate advocate of tripping who died in 2000. Lin was attracted by the message that psychedelics could increase his imagination and bring him closer to Nature.

Lin has a complex story to tell—during one trip he sees himself being fired out of a cannon into the Milky Way—and he ponders "all kinds of issues about what is reality, what is language, what is the self, what is three-dimensional space and time." This echoes what people report from near-death and out-of-body experiences. In all these cases, the self-model stops being the default viewpoint on reality, or, in neurological terms, the DMN.

By releasing the hold of the DMN, psychedelics allow an opening for altering brain function among people with mood disorders and the like. The DMN was only discovered in 2001 by researcher Michael Richie using advanced fMRI scans. For the first time, researchers could see the interconnected parts of the DMN light up when a person was asked, for example, to choose adjectives that described himself. To a neuroscientist, the increased blood flow to the DMN that accompanies this task shows, in effect, that the self is a brain function—your DMN knows who you are because it created your default responses when you think about yourself.

The increase and decrease of blood flow is a useful indicator in brain research, as are patterns of electrical activity, but these are still not a measure of the mind, only its physical mirror. Psychedelics change the pattern of blood flow in unusual ways. Pollan cites a study in which nineteen subjects suffering from depression improved after taking psilocybin, the active ingredient in "magic mushrooms," after being resistant to conventional drug therapy. The researchers found that blood flow to the amygdala (the brain's emotional center) decreased, but this doesn't explain why the subjects also reported profound mystical experiences.

Many are skeptical about such findings. One problem is that there's no way to give a placebo to a control group that would simulate the real drug's hallucinogenic effects. (No one actually knows why our bodies evolved to have receptors for psychedelics, given how rarely they must have been encountered by our remote ancestors.) But the main research on psychedelics seems to indicate that these substances reduce DMN activity, shutting it down temporarily. In tandem with this shutting down, the person taking the drug feels his ego dissolve, together with the normal experience of mind, body, and world. This underscores the point that mind, body, brain, and world are highly malleable in human perception. We are entangled in the process of holding them intact every waking moment, and psychedelics abruptly and radically loosen our grip.

For Pollan, the experience was a revelation. He came to recognize "the tenuousness and relativity of my own default consciousness." His thorough, objective investigation, which involved taking a gamut of mind-altering drugs but also delving into every nook and cranny of medical research, led him to hope that their use would extend to normal healthy people (although he retained a sensible fear of bad trips). He wanted the psychedelic journey to be considered more than a "drug experience." In his view, the first step should be obtaining the guidance of someone who could explain the meaning of a trip after it was over, if it had a meaning (not everyone agrees that it does).

· · · · · · · · · ·

Where does that leave us? No one really knows, and respected scientific journals routinely turn down research papers on psychedelics. Pollan's book should be read to get the complete story about the future promise of these substances. Particularly intriguing is the use of "micro-dosing," in which tiny amounts of a psychedelic are taken on an everyday basis, enough to loosen the grip of the DMN but not altering normal thinking. The hope is that, instead of altering the mind in a drastic, trippy way, micro-dosing will permit self-awareness to observe and reflect upon new possibilities that a person cannot access under normal conditions.

There's no doubt, from a meta viewpoint, that psychedelics weaken or dissolve the mental constructs that keep virtual reality intact. Some people feel that tripping has tapped into Mind at Large. This sounded highly desirable to Huxley, who was an early champion of guided psychedelic trips into expanded consciousness; this was his ultimate goal. To become respectable, psychedelics had to get past their hippie image, however, and brain scans proved to be the key. But the respectability lent by brain scans also contains a flaw.

To a neuroscientist, the DMN is like the adult in the room, a brain region that keeps our wilder, more primitive impulses at bay. Thus a specialized group of cells has taken over the exact function that Freud assigned to the ego. It is typical of our age that psychiatry is highly dependent on pharmaceuticals to combat anxiety and depression, replacing years of expensive, time-consuming talk therapy. As psychiatry has become a matter of shifting around molecules in the brain, so has everything else about the mind. But the notion that "brain creates mind" is just as fallacious here as elsewhere.

The fallacy crops up glaringly in the DMN, because if it indeed controls the balancing act that is the adult mind, who gave it such ability? How did it learn about the benefits of adulthood in the first place? The first answer that comes to mind is that the DMN arose as a survival mechanism, but there's no proof of this—it's just a blanket generalization from the stock-in-trade of Darwinians. Countless people function well in society without bothering to grow into mature adults.

Researchers treat the DMN as if it were a conscious agent with flexible intentions and good judgment. Attributing such qualities to brain cells is a form of magical thinking. Clumps of chemicals don't understand how life works; only consciousness does.

After trying a number of different drugs, Michael Pollan had his most profound experience without them. He was shown how to go into a trance state simply through breathing rapidly and listening to rhythmic drumming. His reaction: "Where in the world has *that* been all my life?" It was in Mind at Large.

All drugs have side effects, some of them unpredictable, but the pitfalls of psychedelics are unique. If you go inside and mess around with a piano's strings, music starts to come out sounding distorted. To go into the most sensitive areas of the higher brain incurs a similar risk, but my point isn't to sound alarms grounded in fear and suspicion—quite the opposite. Mind at Large contains the full range of human potential, and it can be accessed naturally through yoga, meditation, and various contemplative practices. These practices also bring beneficial brain changes, accomplishing this through the most natural of mechanisms: the mind learning to know itself.

The time for psychedelics to come out of the shadows is now, giving us a balanced picture of what is at stake and also at risk. The sudden feeling that we can fly turns out to be a fatal delusion if we test it by jumping off a high building, as has happened more than once. The insights that dawn during a trip often turn out to be unintelligible on returning to a normal waking state. The bigger picture goes beyond psychedelics. Pollan encourages "neural diversity" in the epilogue to his book. But, for me, the mind-altering effect of LSD bears fruit only if it leads to self-awareness. A drug can show you what lies beyond the limited mind; only self-awareness allows you to inhabit Mind at Large permanently.

The opposite of self-awareness is the mechanistic view that a human being is a brain puppet, a subordinate of neural activity. Insofar as psychedelics have medical uses, we should cheer them on. But Huxley's insight that Mind at Large is the real issue still holds true.

What separates us from Mind at Large—or metareality—is the thinnest of veils. It takes only the merest thought. I found this out recently when I was doing the tree pose with a yoga instructor. The tree pose requires delicate balance on one foot, and I've gotten better at it, but suddenly that morning I began to wobble. Immediately the instructor said, "What were you thinking about?" It was the right question to ask, because I had lost my clear, open state of mind as a result of a distracting thought (about the medical benefits of psychedelics, as it happens). It takes a clear, open state of mind, uncluttered by random thoughts, to do a lot more in life than holding a yoga pose. Such clarity is what Mind at Large feels like. It is what freedom feels like. It is the door to metahuman.

In Part Two, we'll make everything personal by showing how each person can experience the shift to Mind at Large. A collective awakening is made possible only when individuals wake up—and nothing is more urgent. As a bridge, let me underscore how extraordinary waking up is, and how unpredictable.

Waking up makes us more human and more real at the same time, because we are the species of consciousness destined—and designed—to know our source completely. If a single vision can unite the human race, this is it.

PART TWO

WAKING UP

# 7

# PUTTING EXPERIENCE FIRST

The punch line to an old joke has gotten embedded in popular culture. As I first heard it, the joke involves a lost city slicker stopping his car on a country lane to ask a farmer how to get to a certain town. The farmer scratches his head with a bewildered look and says, "Sorry, mister, but you can't get there from here."

Why does the joke make us laugh? The humor is rooted in the fact that you can get anywhere on the map from anywhere else. But if the city slicker had asked the farmer how to get to metareality, he'd have gotten the same answer, only this time it wouldn't be funny—at least not to me.

You cannot get to metareality by clinging to virtual reality. The reason for this has been expressed in various ways ever since *Homo sapiens* realized that there is a dimension of life beyond the everyday. Being stuck in virtual reality is like being under a hypnotic spell, and we can't break the spell as long as it has us in its grip—there is no hypnotist standing by who can snap his fingers and break the spell for us. Or we are like dreamers captivated by the dream's illusions, and we cannot wake up as long as the dream lasts—there is no one standing beside the bed to shake us awake.

Spells, dreams, enchantments, illusions, wizardry and witchcraft, mischievous gods—every culture has invented versions of the same idea that reality is deceptive. "You can't get there from here" as long as you are entangled in a complete, all-enveloping trick of the mind. Figuring out how to wake up, therefore, has been a mysterious business. Up until now.

Life proceeds by knowledge and experience, and Part One was all about the knowledge of metareality. So where will the experience come from? Why didn't we delve into it first? It's easy to be impatient (the male sex, in particular, is notorious for not reading the manual or asking for directions when lost), but in this case knowledge and experience can't be separated. You can't jump into the deep end of the pool and figure out how to swim. Consciousness must be reframed from the ground up; perceptions must shift; interpretations have to be abandoned. Nothing can remain quite the same in the leap from human to metahuman.

As all-encompassing as virtual reality is, a loophole exists, which is nothing more than experience itself. What is an experience? Sailing the Atlantic in a one-man boat is very different from climbing the Alps, commuting to work, or baking an apple pie, so the face of experience is constantly changing. But at bottom every experience is the same thing— a conscious event. If you make the mistake of rooting your experiences in the physical world, you cannot wake up from the spell/dream/illusion, because the physical world *is* the spell/dream/illusion.

# Experiencing the World

The only way to break the spell is to reverse the normal explanation of how things work. By this I mean putting the *experience* first. The normal explanation is to put *things* first, which makes common sense. Didn't the stars exist before we came along to gaze at them in the night sky? Without the Earth, which existed in a primordial state before life began, we wouldn't be here. Or so common sense dictates.

Reversing the explanation makes just as much sense, however. Let's say it is morning, and you're reading with a cup of coffee beside you. A "cup of coffee" can mean a physical object fixed, for the moment, in time and space. I've put "cup of coffee" in quotation marks because you are actually relating to a so-called object. In reality, you are experiencing a fusion of sensations. Your eyes see color, light, and shade, which you perceive as the cup. Your nose detects the coffee's aroma, your hand feels the warmth of the coffee in the cup, your tongue its taste.

Take away all these experiences and what happens? There is no cup of coffee. This is a simple, logical conclusion, but if you want to reinforce the spell/dream/illusion, this conclusion is outrageous. A cup of coffee is simply there, standing in space-time as a physical thing. We'll leave aside the quantum explanation that dissolves physical things, including the entire universe, into invisible virtual states. Our goal is different here. The reason for making a cup of coffee vanish isn't to prove that it was never there. The reason is to put experience first. What we call a tree, a cloud, a mountain, a star, or a cup of coffee exists *only* as an experience. Once you accept this fact, the path to metahuman is free and clear.

It does little good, however, to shrug your shoulders and say, "All right, I accept." The words are hollow unless there is an "aha," and you get the point. Experience is how we know anything. If there is a reality beyond human awareness, by definition we will never know it. Much news has been made in contemporary physics by "dark" matter and energy (which Menas Kafatos and I discuss at length in *You Are the Universe*). What makes some energy and matter dark, if some theories are right, is that they exist outside the boundaries of the physical universe. In the dark domain, there may be "stuff" that isn't atomic in origin, doesn't emit photons and electrons, and perhaps has no relationship to our version of space and time.

Let's say that all these conjectures are true (there is a chance they aren't, but we won't go into the technical reasons for that here). Wouldn't dark matter and energy, being totally alien to human experience, having nothing in common with the "stuff" that constitutes the human brain, qualify as real but impossible to experience? For that matter, no one can experience the big bang, because it preceded atoms and molecules. No one can dive into a black hole to experience it, since the extreme gravity in and around a black hole disintegrates matter and tears all ordinary "stuff" apart, including time and space.

For anyone who accepts the universe as it appears, putting experience first sounds nonsensical—but it isn't. Dark matter and energy are known through experience, as are the big bang, black holes, and all the other exotica of physics. In all these cases, the experience is indirect, formulated in mathematical equations, the collection of traces of fleeting subatomic particles, and the data received by radio telescopes and the like. But no matter how indirect, *it is still an experience.* Someone has to see something, if only a sheet of paper filled with numbers. Someone has to hear another scientist talking or read her words. The fact that even the most exotic science is occurring as conscious experience is undeniable. It levels the playing field with everyday experience, because a first-grader learning his ABCs is having the same experience as a physicist reading a journal article on quantum gravity.

As a specific species of consciousness, *Homo sapiens* constructed its version of reality from everyday experiences and built everything up from there. We humanized virtual reality according to our experiences. A hot rock in the desert sun that a lizard finds comfortable burns our skin. The darkest midnight frightens us but makes bats feel right at home, just as shivering heights are perfect nesting places for eagles and the ocean is breathable to fish.

But we are not fated to be imprisoned in our experiences. We shape and mold them with extraordinary freedom. The body looks like a fixed object, but we can starve it or fatten it up, build its muscles or let them go to flab. The question is just how malleable the whole "thingness" of things really is. I think of it as a kind of thawing process. When ice jams break up in spring, they don't thaw immediately. There is a transition from a frozen river to a flowing one. The same is true of how we experience the physical world. Instead of being one big, frozen block, it is reducible to the tiniest experiences, like a rock cliff reduced over time by the pounding sea to fine grains of sand. The five senses are exquisitely attuned to fleeting, transient events. I had kept in the back of my mind a small but impressive finding, that the human retina can detect a single photon of light, but recently this isolated piece of data was expanded in quite a startling way.

# The Quantum Becomes Human

Without achieving widespread popular attention through mass media, research has verified that at least four of the five senses are actually capable of experiencing the quantum domain directly, without the use of sophisticated scientific instruments. The basic findings, which come from leading university laboratories, can be briefly summarized.

SIGHT: The human eye, as just mentioned, can detect a single photon. This is the smallest unit of light in the universe, and our ability to detect a photon has inspired researchers to explore whether they can probe the quantum world with the naked eye.

HEARING: The inner ear is so sensitive that it can detect vibrations less than the diameter of an atom. It can distinguish sounds that are only ten millionths of a second apart.

SMELL: It was previously estimated that the human olfactory system could detect ten thousand distinct smells, but the latest research suggests that smell is a quantum sense that can distinguish a trillion different inputs.

**TOUCH:** We can detect tactile sensations down to one billionth of a meter.

**TASTE:** This sense hasn't been traced to the quantum level, but it is already known that the human tongue detects the five tastes (sweet, salty, sour, bitter, and umami, or savory) at the molecular level. Taste requires smell to distinguish at a finer level, so even taste, when combined with smell, is involved in quantum detection.

Why did evolution give us this micro-sensitivity? To answer this question, you have to appreciate the enormous leap that the new research has taken. Previously, the five senses were thought to operate on the molecular level, like taste. To tell the difference between salty and sweet, for example, the receptors in your taste buds are designed to latch on to the specific molecule of each. Physically, the accepted view has been that we interact with the world through a myriad of similar receptors on the outer membrane of cells. These receptors have been described as keyholes into which very specific molecules (the keys) fit. To smell a rose, for example, molecules of its scent drift through the air and fit into the receptors of olfactory cells in the nose. Because this process occurs at the molecular level, even primitive organisms are physically micro-sensitive.

The most ancient sense is touch, which evolved in single-celled organisms that respond to being touched, like the amoeba, as well as some plants; the Venus flytrap closes its jawlike leaves on its prey when an insect touches the sensitive hairs that line the leaf. But despite being the oldest sense, touch is not yet fully understood. Cameras and listening devices can duplicate the sensitivity of our eyes and ears, but we can tell the difference between wood, metal, and glass by touching them, beyond what engineers can do artificially. The fact that our fingers are sensitive down to a single molecule was only recently proved by experimenters at the University of California–San Diego.

They took fifteen people and gave them three silicon wafers to touch, asking them which of the three felt different from the other two. The wafers were identical except for the top layer, which was a single molecule thick. One surface had an oxidized top layer that was mostly oxygen, the other was coated with a Teflon-like substance. Participants were able to pick out the different wafer 71 percent of the time.

Yet even molecules are massive compared with quanta, and the eye's sensitivity to a quantum of light led scientists to consider the other four senses. It now seems that the entire body is a quantum detector. At first sight this seems remarkable—it certainly expands human perception far beyond previous estimates. At a deeper level, however, these findings indicate how we design and control the constructs of virtual reality. Through the body's quantum detectors, we are seamlessly woven into the universe at the finest level. No longer simply receiving raw data from the world "out there," our bodies participate at the crux where mind and matter mingle.

Getting down to the nub of how we physically see, hear, touch, taste, and smell says nothing about what the five senses are for. Earlier on (page 14), I compared the retina's response to light with a Geiger counter clicking every time it is bombarded with beta particles and gamma rays. But a Geiger counter doesn't experience the world as we do. The sensitivity of our quantum detectors indicates how finely calibrated our experiences actually are. Previously, the fact that we are fine-tuned to the molecular level led to remarkable feats of perception before the recent quantum discoveries.

Professional noses in the perfume industry go far beyond the ordinary sense of smell, being able to distinguish dozens of different rose scents, for example, but the experts don't necessarily have super-refined olfactory nerves or more of them than the rest of us. They've trained their awareness of aromas. The same holds true for professional wine tasters in the area of taste and marksmen in the area of sight. These people may start out with more acute senses than the average person, but a wine taster or a marksman doesn't need an overabundance of cell receptors. With a normal range of receptors, wine tasters sharpen their perceptivity, which is mental.

So-called supertasters have thirty taste buds in a specified area of the tongue, compared with average tasters who have between fifteen and thirty taste buds in the same area, while so-called nontasters (a better term would be *dull tasters*) have fewer than fifteen. Average tasters can improve their discernment of wine by using their noses more, paying closer attention, and slowing down to savor the taste. It is certainly relevant that between the ages of forty and sixty we start to lose taste buds, and the remaining ones shrink. But that's not enough to explain why older people often lose interest in eating. There may be a general loss of interest in life, or feeling alone and unwanted. The simple fact that some babies are born without a sense of taste but grow up to have healthy appetites points to the mental component that dominates our sensory existence.

The data derived from measurements of how the eye, ear, or tongue works say nothing about our actual experience: how a sunset looks, how music sounds, and how chocolate tastes. Science is about measuring life in all its *quantities*, large and small. Experience is about life in all its *qualities*, which cannot be counted. The question "How many units of beauty did you experience today?" is nonsensical. Beauty is experienced subjectively, which no one disputes. But every perception is also experienced subjectively, which is where the huge gap occurs that I mentioned above. There is a total mismatch between a measurement of wavelengths of light and the qualities of light, particularly colors.

Color is created in consciousness by crossing the gap between quantity and quality. In this crossing, a magic trick takes place by which quantum vibrations are transformed into not just colors but everything delivered by the five senses. It would be convenient if the new research explained how that trick is done, but it doesn't. Pinpointing the five senses down to the quantum level is like having a dog's sense of hearing, which extends far above human hearing. If you suddenly woke up and heard the world as a dog does, this wouldn't explain music or speech or anything else about hearing as an experience. The ear doesn't hear; the mind does.

But at least we have an important clue about the magic trick that turns quantum events into human experiences. By taking perception down to the quantum level, we can say—or at least strongly suppose—that we live at the very level where nature transforms virtual reality into the solid physical world. Here I'm using *virtual* in a specific way. In physics there is a halfway house where particles are invisible and have no fixed location.

This halfway house was needed because of the famous uncertainty principle, which says that particles are actually the collapse of energy waves. Waves extend infinitely in all directions; particles exist in one place in time and space. A virtual particle bridges the two states. It hasn't quite yet assumed the shape of a particle, but it isn't a wave extending in every direction, either.

It is very important to know, as we now do, that human experience doesn't have to wait for the physical world to appear; we are able to perceive the birth of particles, physically speaking. I'm being quite literal here. For decades the "collapse of the wave" function has been hotly debated in physics. The main point of controversy is that standard quantum theory holds that it takes an observer to cause the collapse. This point rankles some physicists and baffles almost everyone else. It is taken for granted in daily life that observing something is a passive act. "Look but don't touch," we tell our children. Yet at the quantum level, looking is as good as touching. A particle stops being uncertain when an observer is present. The wave function collapses and, voilà, a particle can be detected.

I hope this abbreviated explanation makes sense; going into the details leads to a tangle of complications. Suffice it to say that researchers are seriously considering whether this mysterious effect on the quantum, known as the "observer effect," can be explained through quantum detection. If our eyes are physically interacting with quanta, this helps us to understand that observation was never passive. We have been participating at the place where virtual reality gets an opportunity to create physical reality.

# How the Mind Gives
# Things Their Thingness

The new findings don't explain the magic trick that occurs in the gap between quantity—things to be measured and counted—and qualities, the sights, sounds, tastes, textures, and smells that we experience as human beings. Humanizing the quantum is exciting, but human beings don't experience the world through a microscopic level of differences, even though we theoretically could. Instead, we lump experiences together according to useful concepts like color. Red is red, not every different vibration in the wavelengths of light between 630 and 700 nanometers (a nanometer is one billionth of a meter).

Likewise, sweet is sweet, not the molecular interaction between sugar molecules and receptors on the tongue. We are so used to the clumping process that we don't catch ourselves doing anything. But we are constantly taking transient experiences, cramming them into prearranged slots, turning discontinuity into continuity, and making solid what is actually fluid.

The technical term for what's happening is *reification*—giving immaterial experiences "thingness." So convincing is this transformation that rocks seem solid and heavy when, in fact, your mind reified them—you have created solidity and heaviness in your own awareness. This constitutes another outrageous conclusion to anyone who is out to reinforce and reaffirm the spell/dream/illusion. But you cannot thaw out the "thingness" of the physical world unless you break down the process that created it. I'm hesitant to use any kind of jargon, but we need to delve into how reification works.

The dictionary definition of *reify* is "to make something more concrete or real." The mental image of money gets reified into a dollar bill, which you can fold up and stick in your wallet. "Parenting" gets reified when you decide to have a baby you can hold in your arms. What's earthshaking is that virtual reality owes its existence entirely to reification.

The web of connections that entangles everything in the spell/dream/ illusion with everything else comes down to the mind, because connections are mind-made. No object is actually a physical thing, pure and simple. "Object" and "thing" and "physical" are strands of a mental web.

People find it relatively easy to accept that a piece of paper currency is the reified form of a concept (money), but they balk when they are told that the same is true of body, brain, and universe. The key is to reverse the whole process of reification, bringing physical objects closer to reality. As you now view them, the bones in your arm are solid and fixed; it's hard to accept that they are fluid and malleable. But all physical objects are actually processes in motion. Bones are no exception; they constantly exchange a stream of oxygen and calcium at the molecular level. Each bone cell is an activity of life in motion. If you wear shoes that are ill-fitting, your leg bones will gradually bend to conform to your lopsided gait.

The fact that a process is slow or fast doesn't make a difference to the basic reality that things are processes. Broken bones heal more slowly than a cut finger, but the very fact that healing occurs testifies to the body being a process. Bone cancer is much feared because it is so painful, but in their normal healthy state, nerves connect your bones to the brain and then to the outside world. By clumping together all the fleeting sensations you are having right now, you seem to have a body that is fixed, but in reality your body today isn't the body you had as a newborn, toddler, or adolescent; it isn't even the body you had yesterday or five minutes ago.

It's surprisingly easy, then, to reverse the habit of thinking that reifies the body, turning a myriad of interconnected processes into a thing. This reversal helps to get us back to where creation originates, in consciousness alone. When you master the reversal, you can undo anything in virtual reality by tracing it back to the mind's creative genius. This is what it means to thaw out "thingness": objects are reduced to a level of consciousness where we can begin to experience the creative process. As you come closer to pure consciousness, the thawing process become easier and faster. As "thingness" stops being so stubborn, experience is transformed into something fluid, flexible, and malleable. The reality of the "true self" is critical, because, as things stand, everyone has a self filled with contradictions. The self you have been identifying with keeps you in the spell/dream/illusion. The true self takes you from illusion to reality.

# Three Versions of the Self

The true self lies hidden behind layers of disguise. The layers are so thick that no one can confidently define what the self actually is. "Self" is a convenient fiction, lumping together a jumble of beliefs, experiences, old conditioning, and secondhand opinions. This is a greater problem than you'd imagine. For almost everyone, there is a wide gap between "being yourself" and "knowing yourself." The first is considered desirable. When you are able to be yourself, you feel natural and relaxed, without pretense or defenses. Knowing yourself is a different matter. A century after Freud discovered the subconscious mind, it has been identified with the dark side of human nature. We repress our urges of anger, anxiety, envy, insecurity, and even violence. Of course you cannot get along with other people if you say everything you think or act on your every impulse.

But there's more to it. When the inner world is identified with the dark side, people don't want to look there. They dislike and fear what they find, or might find. We identify with the ego-personality, which presents the self we want the world to see, and we ignore the opportunity to explore what a deeper self might be. In time, countless people actually believe that their ego-personality is their real self. Yet there are two more selves we all possess, and they are nothing to fear. In fact, they are the richest sources of human fulfillment.

The first is the unconscious self. Even though we routinely shove negative emotions and impulses out of sight in the subconscious, the whole story is much more positive. The unconscious self is creative and sensitive. When you walk into a room where two people have been arguing or someone was crying, you silently sense it "in the air." Actually, you are sensing it through your unconscious self. At the level below everyday awareness, you constantly perceive your surroundings. You also have the power of intuition in your unconscious self. You gain "aha" moments when the unconscious self reveals something to you that your conscious mind didn't realize.

As you grow to maturity, you begin to value what is rooted in the unconscious. You feel confident, self-reliant, and sure of what you know. You know how to do certain things—cooking, driving, balancing your checkbook, finding a good restaurant. But at a deeper level you have a settled feeling that is hard to explain. Paying attention to all of your life experiences, your unconscious self distills the essence of your life into the experience of inner fulfillment, which over time becomes a natural part of who you are. You know yourself as a set of values, purposes, and achievements. There are countless people who do not arrive at this stage, and they don't know the experience of inner fulfillment. Nor can it be imparted, because so much happens out of sight in the unconscious.

T. S. Eliot comes to mind once again. In 1925 he wrote "The Hollow Men," a poem that high schoolers used to love to recite, because adolescence is marked by hidden fears. The first verse begins:

*We are the hollow men*
*We are the stuffed men*
*Leaning together*
*Headpiece filled with straw. Alas!*
*Our dried voices, when*
*We whisper together*
*Are quiet and meaningless*
*As wind in dry grass.*

Writing in 1925, Eliot is exploring one of our deepest fears, that life can turn meaningless, haunted by death and nothingness—the ultimate hollowness. In times of great peril and horror, as in the two world wars of the twentieth century, the dread of losing all meaning feels very real. Yet the unconscious self, drawing on reserves of infinite meaning, re-creates the world in a new image, one that is livable despite all terror and horror in the past.

There is yet another self, however, and it is even more valuable—call it the "true self." This is a level of awareness very close to our source in pure consciousness. Pure consciousness is silent and still. It has the potential for mental activity before any activity arises. I identify it with the simplest of all experiences, the pure "I am" of existence. This is the simplest of all experiences because it requires no thought. You know that you exist; that's all. As the still silence of "I am" starts to vibrate into thoughts, images, feelings, and sensations, the first stirrings are very faint and subtle. They are highly fluid and malleable, which is why desires and intentions that come from our deepest source are not distorted by all the cruder desires of the ego. "I want peace" is a subtler, finer desire than "I want a Porsche."

At the level of the true self, all desire for change reaches its goal, because only here is "I am" enough to bring total fulfillment. No external gratification compares with this. It seems strange on the face of it that the mind at its subtlest level should be more fulfilled than the mind at more superficial levels. Chasing worldly desires is what life is all about to the vast majority of people. Being still and quiet, by the same token, feels very uncomfortable for many. "There's nothing to do," we complain. Yet stillness can unlock new realities. The key is that pure consciousness contains infinite resources of creativity, bliss, intelligence, love, and awareness. By living close to the source, you have access to this infinite potential that allows you to be a genuine co-creator of reality.

I've put these three selves—the ego-personality, the unconscious self, and the true self—into separate categories only for the purpose of description. In daily life we call on all three. As awareness rises from its source, any impulse has an unconscious component and eventually an ego component. A common example is friendship that turns to romantic love. Two friends interact largely on an ego level, meaning that they present their social persona to each other. But as friendship deepens, the unconscious reveals itself more intimately, and sometimes, if the two people feel safe enough, the true core of friendship, which is love, reveals itself. That seems to be the final goal, but at the level of the true self, no other person is needed. "I am" already has the quality of love, trembling on the edge of pure consciousness.

What this comes down to is that the self you identify with is simply *the self you are aware of.* There is no fixed self, just as there is no fixed body. Fighting for your place in line at the post office calls up the ego-personality. Feeling tenderness toward a baby calls up the unconscious personality. Feeling that you matter in the great scheme of things calls up the true self.

At a certain point, the true self dominates the scene, and when this change occurs, the world changes as well. The world feels hard, fixed, stubborn, and inflexible if you are all those things. The ego-personality finds it easier to resist than to accept, to hold on rather than to let go. So being hard, fixed, stubborn, and inflexible isn't rare, and seeing the world the same way isn't rare, either. If someone explores more deeply and begins to identify with the unconscious self, then the world seems beautiful, fresh, renewed, and filled with light. This, too, is a reflection of the person's state of awareness. Look at the glowing light in French impressionist paintings and you'll see where such a state of awareness leads painters.

Yet you must go deeper to see the world in complete purity. This state, known generally as "enlightenment," represents direct contact with metareality. Since our daily life is dominated by the ego's desires, needs, and demands, it isn't possible to conceive of what it feels like to experience metareality as a constant state. Let me give a striking example in the thinking and teaching of Krishna Menon, born in 1883 in the South Indian state of Kerala. Although obscure most of his life—Menon died in 1959—he has come to be regarded as extremely important among seekers who want to investigate the experience of enlightenment (as opposed to the veneration of gurus and spiritual teachers, which has a major religious component).

The uncompromising nature of his awareness can be gleaned by reading the answers that Krishna Menon, later known to his followers as Sri Atmananda, gave when questions were posed to him. Time and again he took every subject back to one essential question: What is reality? Here are some succinct examples:

> *On pleasure and pain.* I feel pleasure at one moment and pain at
>     another. But I am changeless all along. Thus my pleasure and
>     pain are not part of my real nature.
> *How are physical objects related to thoughts?* The question arises on
>     the assumption that objects exist independent of thoughts.
>     That is never the case. Without thought, there is no object.
> *Should we follow a personal God?* I say no, because a personal God
>     is nothing but a concept. Truth is beyond all concepts.

Such unshakable certainty feels liberating and audacious at the same time. But it's clear that Menon was not speaking merely to draw attention to himself; he was speaking from the perspective of the true self, once the ego-personality and the fragmented mind it creates have been abandoned. This is made evident in another question and answer.

Q: Is reality indivisible?

A: Reality is only one and cannot be affected by quality or degree in any way. Reality is purely subjective. I am the only subject, and all the rest are objects. Diversity can be diversity only through me, the "One."

You have to reach the end of this answer to discover that far from being an expression of solipsism ("I am the only subject, and all the rest are objects"), Menon's viewpoint reflects his sense of wholeness (the One). This way of speaking echoes an Indian tradition going back thousands of years, known as *Advaita*, which in Sanskrit means "not-two." One could also call it the perspective of the whole mind, because the primary purpose of Advaita (and metahuman) is to get people to identify with consciousness as a whole, instead of its fragmented products.

I've come to the point where I hesitate to adopt spiritual language, preferring to see the inner journey as an escape from illusion to reality. The question still lingers: If Menon, along with others who are called enlightened, are so exceptional, do they stand too far outside the norm? I say no. The process of waking up is natural; anyone can do it. The proof of this is right before our eyes. Every day we follow the guidance of the ego-personality, the unconscious self, and the true self in a kind of jumble. But the very fact that the true self is speaking to us in moments of love, joy, creativity, and renewal, even if only in fits and starts, indicates that we are in touch with metareality. Knowing that the connection exists, the whole process of waking up can be explored in an organized way. How this is done is the subject of the next chapter in our journey.

# 8

---

# GOING BEYOND
# ALL STORIES

If metareality is "everywhere, always, and everything," a startling fact follows. There is no story we can tell about it. All the elements of a story—a beginning, a plot, various characters, and an ending—have no bearing. Yet everything else in the world is organized around these elements. You are a character in your own story, which had birth as its beginning, death as its ending, and all kinds of plot twists and incidental characters along the way. The prospect of giving up your story is all but unthinkable—yet it is totally necessary in order to experience metareality. We even have to give up mystical, religious, and spiritual stories, too, because they turn pure consciousness into something it isn't, whether that something is the Old Testament God or Nirvana or a pantheon of gods and goddesses. I appreciate that these have been a guiding light for centuries. Isolated messengers have always existed to point to another world, bringing inspiring stories with them.

In his depictions of the Madonna and Child, Leonardo da Vinci followed the traditional story that St. John as a boy was a childhood companion of Jesus. Leonardo (and other Renaissance painters) show him

enigmatically pointing upward with one finger, a smile playing beatifically across his face. *Heaven—can't you see it? Just there,* the smile says.

Those who wake up see the transcendent world directly. The time
frame of eternity becomes natural and is felt as a seamless flow. The Buddha once said (in a translation by Sogyal Rinpoche), "This existence of
ours is as transient as autumn clouds. To watch the birth and death of beings is like looking at the movements of a dance. A lifetime is like a flash
of lightning in the sky, rushing by like a torrent down a steep mountain."

In every spiritual tradition, such messages have beckoned believers to
a transcendent world, but all have failed to convince the average person
that going beyond should be the focus of daily life. At no point in history did waking up go viral. Somewhere in the evolution of consciousness, *Homo sapiens* came to a fork in the road. Collectively, we could have
identified with the true self or else we could have identified with "I," the
ego-personality. Obviously, we took the second road. Metareality didn't
abandon us; we abandoned it.

This made all the difference in how we see ourselves. The true self is
connected to its source in pure consciousness. The ego-personality has its
source only in the stories it imagines and believes in. Modern people have
jettisoned mythology, and many are rejecting organized religion. But at
every level, our lives are still shaped by stories invented by human imagination. The other road, which led to the true self, gained a reputation
for being mystical (i.e., detached from real life), and once this happened,
only a motley crew of saints, sages, poets, artists, and seers followed that
path.

Finding metareality is an impossible task until you question your own
story. You must take the challenge personally, because becoming metahuman is real only if it is real *for you*. If you are fully aware of what is
happening to you here and now, you have gone beyond all stories. The
habit of constantly adding to your story is just that, a habit. By itself, the
present moment has no story—it just is. Why do we embellish it with our
story? Because the present moment isn't fulfilling by itself until it gets

enriched by the true self. A computer is useless until you plug it in, and although we use our minds for everything already, much time is lost in fantasy, distractions, avoidance, denial, procrastination, self-judgment, and on and on. Every story contains these undesirable elements. Staying connected to the true self at every moment brings the fullness of life into our awareness.

Metahuman could be called a new story, another fiction added to the shelf, but I think that would be a misnomer. The element of fiction is taken out when you wake up. As part of the spell/dream/illusion, our personal stories can't do without the fictional element. Virtual reality is a fiction to begin with. Anything based on it participates in unreality. We might pity our ancestors for their attachment to myths, superstition, and unproven religious beliefs. But if we are immersed in a better story, it's still a story. Future generations won't buy into it any more than we buy into Zeus, witches, and the heart as the seat of intelligence rather than the brain (a persistent belief in ancient Greek and Roman medicine).

This chapter is about a turning point where we stop telling ourselves stories, no longer needing them to defend us against harsh reality or to make sense out of a chaotic world. In metareality you go beyond danger and disorder. The true self gives you an unshakable stance in reality, at which point your life finds purpose and meaning from the source instead of from a cobbled-together fiction.

# Clinging to Stories

Humans are proud to be storytellers, and our story has a gaping hole because no record was left from the oldest times, before writing emerged. Can we imagine so many lost stories? Around 45,000 years ago, people we'd recognize as modern *Homo sapiens* started to migrate northward from Africa. All were hunter-gatherers. Many generations before farming, mining for metals, and fixed dwellings heralded the rise of civilization, life 45,000 years ago was already too complex to do without stories. In the human mind, fire had to come from somewhere, rain had to be unpredictable for a reason, and things we now take for granted, like a chick hatching out of an egg, posed a deep mystery. Mythology arose not as fantasy but as the best way to explain Nature, given the life people were leading. Ascribing meaning to anything and everything is the thread that binds us to earliest humans. Stories explain how life works and thus fulfill a need that is woven into the fabric of being human.

We still live by stories, and whatever disrupts our personal story is usually rejected out of hand or fought against vigorously. (Witness the guiltiest sexual abusers coming to light through the #MeToo movement, abusers who flatly deny any wrongdoing.) We've already discussed how the ego creates the illusion of being separate and alone. "I" needs a good story to feel safe, important, socially acceptable, and worthy. Seeking to feel secure, people want to belong to something larger than themselves—a tribe, a religion, a race, a nation—but to be accepted by any of these groups, you first must accept their story. Without thinking about the freedom they are surrendering, people adopt a secondhand, sealed-off story. You know instantly who is "us" and who is "them." Yet no matter how convincing the story, you will always be "them" to people embedded in another story. The security to be found in something larger than yourself breaks down when "they" turn into a threat, even an enemy of your very survival.

Stories grow out of such basic needs that going beyond any story feels impossible. A story consists of anything the human mind can imagine, which leads to infinite choices. But we can simplify the matter. Stories are about attachment. We think to ourselves, "I am X," and then we cling to X as part of our identity.

X can be the larger group (tribe, race, nation, religion) mentioned above. "I am American" has enormous power over people, as does "I am French" or "I am Jewish" or "I am white." But in seeing this, which most people do, we are only scratching the surface. Any version of "I am X" can lead to clinging and attachment. "I am a Patriots fan" or "I am upper middle class" creates passionate attachments. At the same time, stories become more potent by what they exclude. For every "I am X," there are many possibilities of "I am not Y." If you are American, you are not every other nationality, of which there are hundreds. If you are Catholic, you exclude all other faiths, and so on.

What's wrong with this? If stories were held lightly, the way we experience *The Hobbit* or *The Great Gatsby*, being diverted for a while before moving on, vexing problems wouldn't arise. It's not the story that is responsible; it's our attachment to it. Attachment falsifies experience by freezing it in place. The weight of the past becomes a burden. The present moment is lost in the clutter of memory, belief, and old conditioning. How many older people crave to be young again? How many regrets do we harbor that we refuse to let go of? These attachments exist in everyone's life, leaving aside the misery and violence created by an us-versus-them mentality.

Prying someone loose from a cherished story lies at the heart of one of the most popular best sellers in the 1980s, M. Scott Peck's *The Road Less Traveled*. The book begins with a gripping sentence: "Life is difficult." Immediately, Peck expands on what he means: "This is a great truth, one of the greatest truths. It is a great truth because once we truly see this truth, we transcend it . . . [and] the fact that life is difficult no longer matters."

In other words, we are hearing the call to go beyond. But Peck, who was a psychiatrist, knew that just because a truth is great doesn't mean people can face it—quite the opposite. In his view, after years of treating patients in therapy, the biggest obstacle to getting better his patients faced was their refusal to take responsibility. Why? "We cannot solve life's problems except by solving them," Peck writes. Yet this fact "is seemingly beyond the comprehension of much of the human race. . . .

Many, so many, seek to avoid the pain of their problems by saying to themselves: 'This problem was caused by other people, or by social circumstances beyond my control, and therefore it is up to other people to solve this problem for me.'"

Peck gives vivid examples to illustrate his point, such as the case of a woman who had just tried to slash her wrists. She was an army wife on the Pacific island of Okinawa, where the younger Peck was an army psychiatrist.

In the emergency room he asked the woman, who had managed to inflict only superficial cuts on herself, why she wanted to commit suicide. She said it all came down to living "on this dumb island." Not satisfied with her answer, Peck began a dialogue that would be comical if the woman weren't so miserable and unhappy. The back-and-forth between him and his patient can be summarized as follows:

PECK: Why do you find being on Okinawa so painful?

PATIENT: I have no friends, and I'm alone all the time.

PECK: Why don't you have any friends?

PATIENT: I live off base in the village, and nobody speaks English.

PECK: Why don't you drive to the American base to find some friends?

PATIENT: My husband needs the car to drive to work.

PECK: You could drive him to work.

PATIENT: It's a stick shift. I only know how to drive an automatic.

PECK: You could learn to drive a stick shift.

PATIENT: Are you crazy? Not on these roads.

What made the scene tragic and comical at the same time is that the woman, despite her absurd stubbornness, was in genuine pain; she was crying during most of their conversation. We all believe in our story, almost at any cost. From a therapist's perspective this woman was skirting any hope of getting better by exhibiting all the classic signs of denial.

Who among us hasn't shut our eyes when a situation was simply too hard to face? Life would be much easier if our problems had cut-and-dried solutions, but they don't.

Denial is only one element in the big picture, which is that we live by stories. Denial comes down to ignoring anything that disrupts your story or contradicts it. Even the healthiest person psychologically ignores an enormous part of reality. If we see, once and for all, that being attached to stories is what we want to overcome, we recognize that there are people who have done just that. Their inner world isn't cluttered with the baggage of the past. They don't defend "I am X" as if their survival depended on it. Rather, they live effortlessly in the present. The real issue isn't whether such people exist; the real issue is why we have ignored them for so long.

# The Awakened

States of awareness long considered mystical in the West—if not bogus—exist all around us, affecting ordinary people who almost always keep their experiences private. A few years ago I became acquainted with Dr. Jeffery Martin, a PhD social researcher who had conducted groundbreaking research into higher consciousness. What made his studies revolutionary is a single basic discovery: many more people are awake than anyone ever supposed.

Martin gave them a voice. After gaining his doctorate at Harvard's School of Education, he started publishing his findings, which indicated the prevalence of enlightenment as a natural state of awareness, one that many people have already accessed. Martin's research began with him posting an online query asking for responses from people who thought they were enlightened.

To his surprise, he got more than 2,500 replies, and from this pool Martin intensively interviewed around 50 subjects. At first it was hard to find a common language. Feeling that you are enlightened is personal, and it also sets you apart from normal society. Martin's subjects were sensitized by being outsiders, often being stigmatized in adolescence for being different. They also knew on the inside that they weren't normal according to the social standards around them. For many, revealing their unusual mental state had led to such things as being sent to a psychiatrist, put on medication, or even being committed to a mental hospital.

However, very early on, Martin realized that as different as each of his subjects was, their experiences fell on a continuum. There wasn't a single enlightened state, but rather a sliding scale. To find some common ground, and to fit the accepted model of what doctoral research in psychology must look like, he dropped the loaded term *enlightenment*, and in its place adopted the cumbersome tag *persistent non-symbolic experience*. When someone begins to have such experiences, "[t]here is a shift in what it feels like to be you," Martin notes. "You move away from an individual sense of self, which is considered normal, to something else."

Defining what that "something else" is wasn't easy, because these people came from different backgrounds and were influenced by diverse cultural factors. However, Martin was able to identify specific areas that seemed to stand out as markers of higher consciousness.

Typically these people said they had lost a sense of being a separate self—they had no lingering notion of personal identity. Putting himself in his subjects' shoes, Martin comments, "I would be saying, 'There's no Jeffery here talking to you.' That's literally what they would say to me."

Another common experience was a dramatic reduction in thinking. "In fact," Martin says, "they would often report having no thoughts." This wasn't literally true, as Martin discovered when he did further research, but as a self-report, having no thoughts at all is startling. Another common experience was a sense of unity, oneness, and wholeness. This state of completeness, Martin says, led to a tremendous sense of personal freedom. "There's a loss of fear that comes with this, a loss of identification with a personal story." Many felt that their bodies were no longer bounded by the skin but extended beyond the physical body.

So now we have an objective profile of shifting into higher conscious-ness in everyday life, the state of metahuman. Personal awakening isn't rare, and according to those who have actually experienced it, their awak-ening led to a wide range of possibilities. The implications for human nature are intriguing, beginning with what it feels like not to defend "I, me, and mine." The source of so much anger, fear, greed, and jealousy is rooted in the insecurity of the ego and its endless demands.

Putting ourselves in their shoes, these awakened people don't keep a running story in their heads about what's happening to "me." When they think about themselves, "me" fades away as soon as they notice it. The same is true for their emotions, which are fewer and more sponta-neous. When anger arises, it fades almost immediately. Emotions were still positive and negative but rarely, if ever, were they extreme. Martin's subjects could be irritated when something bad happened at work, yet they didn't carry the residue of stress around with them afterward, and it would never build into angry frustration. They felt a sense of inner peace that could be interrupted, but quite soon it would return. In a word, these people were very good at letting go.

To bring order to his findings, Martin divided the awakened state into several "locations" as separate stages of awakening, according to their intensity. His subjects regarded waking up as a definite shift, which for some occurred as recently as six months previously, for others as long as forty years before. Once they arrived in Location 1, as Martin calls the opening stage of awakening, people usually kept progressing and rarely slipped back or jumped ahead. In other words, they were experiencing personal evolution, and the process showed no prospect of stopping. As Martin described their attitude, they "agreed that their initial transition was just the beginning of a process that seemed to be able to unfold, and deepen, endlessly—a never-ending adventure." Everything happened internally, and for many the shift wasn't something they would define as spiritual. It was just the way they experienced themselves.

"If you sat in a roomful of people and a small percentage belong to this altered sense of self," says Martin, "you wouldn't be able to spot them. To all outer appearances they are just like you and me."

So who exactly are these people? At the outset, before Martin's research began to expand to many universities and countries, his typical subject was a white male from the United States or Europe. Martin was disappointed to find that women, for unknown reasons, were not eager to volunteer as enlightened or to discuss their experiences of "waking up." Religious backgrounds were diverse, spanning Eastern and Western faiths, yet most of the subjects had done some kind of spiritual practice—they wanted to be in a higher state of consciousness. Curiously, around 14 percent had done nothing of this kind. They had spontaneously popped into higher consciousness or, more typically, drifted into it.

Martin's research base has expanded to more than a thousand subjects, which means we must ask ourselves if "normal" awareness isn't a fixed state at all but a spectrum, with consciousness evolving much further than anyone has previously predicted. At the very least, higher consciousness has become much less exotic. It's no longer the province of sadhus and yogis in the Himalayas.

When we communicated by email and later met in person, I accepted why Martin, for academic reasons, had to remain value-neutral. But waking up has for centuries been associated with bliss, for example, or communication with higher beings like angels, or feeling a divine presence. Was all of that absent when modern people woke up? Martin assured me that "outside" phenomena were present, but he felt constrained not to mention them in his doctoral thesis. He found that the spiritual dimension of some of his subjects had also opened up. Some of the people he tested reported the kind of open, clear, silent awareness associated with Buddhism. Still others, however, had no idea what to make of their state of awareness.

In later reports Martin notes that a small number of his subjects had experiences that defy normal explanations. A small number "experience a deep bliss sensation all through their body, including during moments that would otherwise be physically painful. For some this seems to bring pain tolerances that appear infinite. A few of [them] have reported experiences that should have involved horrific amounts of pain, but only resulted in bliss. Others who experience ongoing bliss find that they can reach its limits. They report a threshold, unique to each individual, above which pain is experienced."

It turns out, as his researches deepened, that Martin discovered more and more uniqueness among these people. Beyond Location 4 a few progressed as far as Location 9, for example. In that location, Martin notes, people would "say something like 'it feels like it is just the universe looking out these eyes.'" But generally speaking, all of his subjects were amazed at the amount of well-being they were experiencing, and this grew as they moved further along the spectrum. Paradoxically, in Location 4, all emotion fell away, even love, which Martin associates with a larger shift—the self that was built up through a continuous story (he calls this the Narrative-Self) falls away, along with socially defined emotions. After Location 4, emotions begin to return in a different form, based on the foundation of continuous well-being. Yet even before this return, his subjects reported that they didn't miss the experience of emotions, because in freedom they had found the highest state of well-being.

# Teaching People to Wake Up

Martin had made what he considered "the fundamental discovery that these were psychological states that had been identified and adopted for thousands of years by many cultures and belief systems." In the present climate, where consciousness is an exploding growth field, the real issue isn't skepticism. In one survey, one-third of American adults believe in things widely considered fringe or New Age, from reincarnation and the paranormal to medical applications long opposed by mainstream medicine. (According to various sources, between one-third and 38 percent of American adults use alternative medicine. This includes the 30 million, for example, who visit chiropractors every year.)

Martin was not unique in saying that higher consciousness wasn't "inherently spiritual or religious, or limited to any given culture or population." Given his academic and technical bent, he decided to put his data to work. He sifted out the techniques that his subjects considered the most powerful in getting them to arrive at higher consciousness, and he organized these into a fifteen-week Finders course. What's fascinating is that the students would be ordinary people who, for whatever reason, were attracted to take the course.

Three participants gathered at the website Reality Sandwich to report on their experience: Catherine, a business and leadership consultant from Paris; Paul, the co-owner and manager of a garden center in Wales; and Rebekah, a semiretired photographer from Texas.

Each had different reasons for taking the course. Paul described a period of personal difficulties. "I'd become disconnected, disillusioned, mostly due to physical and material matters. I had a very spiritual upbringing, but I seemed to have lost the plot altogether." His general state was "not quite suicidal, but about as low as I could possibly be."

Catherine had heard about Martin's research and was intrigued by the possibility of higher states of consciousness. It particularly interested her that these states could be specially described—"It's not just nirvana all day long. My objective was to go through the experience and to reach higher states of well-being and quiet."

Rebekah had no prior expectations. "[I] did not know what to expect, just open to whatever." But she had heard about Martin's research projects, and said, "I trusted the science in them." She also knew what she wanted out of the course. "My goal was spiritual evolution. How can I raise my consciousness to a higher level?"

The methods presented to them were highly intensive, amounting to two to three hours a day, which they were expected to keep up when they went back to their everyday lives. The instruction consisted of a weekly video about what the participants would be doing for the following week. Before the next video, as Catherine describes it, "[y]ou do a summary of the week. How do you feel? What has happened to you? How many times a day did you do the different activities?" There was meditation and also group discussion. Some exercises came straight from standard therapy, such as writing down and forgiving people from your past who had wronged you.

At the core of the course, however, was Martin's eight-year research into people who considered themselves enlightened. He gave each a questionnaire about which practices they found most useful in their journey, leading to a wealth of data. "We looked at all of that, and only a handful of things rose to the top. Some of them were across all traditions, like, for instance, a mantra-based meditation practice." Other techniques were more specific. For example, says Martin, he adopted a "direct awareness–type method. This involves placing your attention on awareness itself. Now that sounds simple, but as I'm sure all of these folks will tell you, it's rather tricky."

From students' feedback, he quickly learned that some practices worked better than others for each person. He also mixed in techniques: "It's not all ancient practices. We also include some of the gold standard exercises from positive psychology." In a general way, the course had two aims, to increase well-being as quickly as possible and to deepen awareness. The success that resulted sounds remarkable: "More than 70 percent of participants who completed the course report having persistent forms of an 'awakening' experience, and 100 percent say that they are happier than they were when the course began, even those who rated themselves as 'very unhappy' at the start of the course."

Paul, the near-suicidal manager of a garden center, offers personal testimony to that: "It was the dropping away of sloth; that was the biggest single thing that I noticed. The chatter was going, the general day-to-day worries and anxieties were dropping away at a terrific rate. That was what resonated for me initially. The lack of fear, of worry, of anxiety. That was the biggest impact for me initially."

So do we have a final answer—has metahuman become accessible through a cocktail of psychotherapy, group sharing, self-help, meditation, and a program as personalized as a fitness regimen at the gym? There's no cut-and-dried answer. Too many people find too many paths to waking up. Martin's research is only one version of a trend to make the mind a kind of techie project, and although he claims that 70 percent of participants who experienced some form of awakening continued to have the experience, only time will tell. An intensive lifestyle change that includes hours of practice a day will attract only the dedicated few.

The mystery of awakening includes the 14 percent in Martin's surveys who awoke spontaneously. One day, without warning, they found themselves fully self-aware or they eased into that state over time but without effort. We've already touched on a similar phenomenon in sudden genius syndrome (see page 80), and there are rare cases where people discover suddenly that they have total recall of everything that has happened in their life (a phenomenon known as "superior autobiographical memory"). Such people can get together and chat about things like "What was the best Tuesday in your life?" or recall the theme song to a TV series that aired only a few times in the seventies.

In all these cases, a person's awareness doesn't accept imposed limitations as normal. Under hypnosis, ordinary people can uncover detailed memories they can't otherwise retrieve, such as knowing the number of rosebushes in a garden from childhood or how many stairs led to the basement in their parents' home. Is it normal to remember or forget? Both, of course. Unfiltered raw data bombards us in waves too overwhelming to absorb, so we selectively forget and remember. The point of waking up is to remove some barriers created by memory and other barriers created by forgetting. Happiness is blocked if you keep remembering and rehashing old hurts, but just as effectively if you forget how happy you once were—it's a matter of perspective. You can even say that virtual reality makes us forget to remember who we really are.

At its most universal, enlightenment is simply expanded self-awareness. We go beyond stories, beyond fixed boundaries, beyond the rickety construct of "I," and, in doing so, awareness effortlessly expands. It expands naturally, of its own accord, because stories, boundaries, and limitations were artificial to begin with.

# 9

## THE DIRECT PATH

The guides of humanity who urge us to undergo radical change use many tactics, including the religious promise of eternal peace and unbounded happiness, either here or in the afterlife. The carrots lure us on, but there's also the stick—the persistence of present suffering and fear of more suffering to come. The stick doesn't work in getting people to change, because even the ultimate threat, unending suffering in hellfire and brimstone, can't compete with the urges of desire and the reckless behavior that follows. The carrot isn't much better. Why should we believe in divine rewards? Anyone can see that virtue very often is not rewarded while sin is. Soldiers are told that God is on their side, which gives divine vindication for going to war. But, at the same time, urging soldiers to wipe out the enemy contradicts the teaching that God forbids killing.

When God is used to justify killing and condemn it at the same time, we are almost too deep in the spell/dream/illusion to be rescued. Every religion has tried to make the rewards of salvation, redemption, enlightenment, waking up, or whatever you want to call it more extravagant.

Consider the Parable of the Burning House, which appears in an ancient Buddhist scripture, the Lotus Sutra.

A rich man's house is on fire and, to his dismay, his children refuse to leave. They are so absorbed in playing with their toys that they ignore the flames all around them. Frantically, the rich man gropes for a solution, and one occurs to him. He tells his children that waiting outside the house are even better toys. Their fondest wish was for a pretty goat-drawn cart, and one is right outside the door. Enticed by their father's promise, the children run out of the burning house, and what they find isn't an ordinary goat-drawn cart, which would have made them happy, but a far more splendid reward, a jewel-encrusted cart pulled by two snow-white bullocks.

This carrot, like the promise of a heavenly feast prepared by God with Christ sitting at the head of the table, demonstrates how far religions will go to win adherents. But, for many, these inflated promises have lost their power to inspire devotion. I don't think modern secular people are less faithful—they've simply shifted their faith from religion to science. The spell/dream/illusion keeps changing. In prosperous societies, most people live in an upgraded illusion filled with the wonders of technology. Why should they give that up? In sheer exasperation, the great Persian poet Rumi looks at humanity and pleads, "Why do you stay in prison when the door is wide open!" The honest answer is that we like it here. It took enormous creativity to devise the current state of our collective dream. But the key isn't wealth and comfort. At a very deep level, humans agreed to a life of high drama. The drama is driven by the warring opposites of pain and pleasure, desire and repulsion, good and evil, light and darkness, us versus them, my God against your God, and so on.

There's no sign that the drama will ever end. The direst threats, like global climate change, only exacerbate it. As for new carrots, I doubt that waking up can compete with going to Heaven or finding a jewel-encrusted cart. There has to be a way to reach metahuman that isn't either carrot or stick. Such a way does exist; it's known as the "direct path." It offers no punishment and no reward. What else is there? Wholeness. This answer doesn't seem very enticing at first. We're too used to the system of rewards and punishments meted out by the war of opposites. But wholeness is healing—it overcomes the wounds of separation and the divided self. Wholeness is unshakable and eternal. It is the only thing that no one and nothing can take away from you, not even death.

I don't want to make these things seem like another set of rewards. The direct path isn't a sneaky tactic for dressing up Heaven in a new costume. Reality is whole already, needing no human beings to validate it. Wholeness is totally real. The direct path exists to show the way to the same reality. If it succeeds, the outcome will be unpredictable. Reality is confronted here and now, in the midst of constant change. The here and now can't be pictured or described. But by ending the war of opposites, wholeness erases a false reality—no reward we can possibly imagine compares with this.

The direct path, also known as the direct method, aims to shift a person's sense of self. Instead of "I am X," which forces us to identify with bits and pieces of experience, we become content with "I am." This is more than a semantic twist. "I am" means that you identify with existence, and because existence contains the infinite potential of consciousness, so do you. The shift to "I am" involves every aspect of life, and this poses a potential problem. Which aspect should we confront first—the body, the mind, the brain, psychology, relationships, beliefs, or habits and old conditioning? It's not clear that any of these should be given top priority, or confronted at all.

The direct path doesn't ask us to think about our problems intellectually or to analyze the prospects for inner change. In fact, the direct path bypasses all ordinary ways of seeing and thinking about ourselves. These are what got us into trouble to begin with. Ordinary ways of seeing and thinking are totally adapted to virtual reality. A different way must be found.

In my experience, *direct* is a tricky word. It implies something that happens immediately, but going to college for four years is the direct way to get a bachelor's degree, and four years isn't immediate. *Direct* also implies something easy, efficient, without obstacles. This applies, for example, when a package is delivered *directly* to your house. You don't have to go to the trouble and expense of driving to the manufacturer to get whatever is inside the package. But a Sherpa leading a trek to the top of Mt. Everest is exactly the opposite—the path might be direct, but it is arduous and full of obstacles.

The origins of the direct path can be traced back to Vedic India and ancient Greece, where the issue of waking up was fully discussed. Many answers and many ways to reach the goal were devised. A detailed survey of all these paths would lead to more confusion than ever, because disagreements among the various approaches abound. There have been dreams of shortcuts, such as accepting Christ as the Son of God, a single decision that redeems all sins and opens the door to Heaven. That's as direct as it gets. On the other hand, a lifetime can be spent by a Tibetan Buddhist monk or a devout Hindu renouncing the world for a forest ashram or a mountain cave. This choice presents the prospect of a long, difficult inward journey.

In this book I've been describing the simplest and most powerful (I believe) version of a shortcut, which is to exchange illusion for reality. I define *direct* as easy, efficient, and natural. There is no need for anything arduous and filled with obstacles. I insist on a path that is painless, because I've seen too much of the opposite. Some people spend years in frustration and disappointment struggling to reach some spiritual goal that remains just out of reach. What baffles people is that the whole spiritual project gets tangled up in misguided ideas and doomed expectations. Let me list the pitfalls one is most likely to encounter:

1. Mistaking the goal for some kind of self-improvement, shedding your old imperfect self for a shiny new one.

2. Assuming that you already know what the goal is.
3. Hoping that higher consciousness will solve all your problems.
4. Struggling and striving to get somewhere fast.
5. Following a cut-and-dried method, usually a method backed by some famous spiritual authority.
6. Hoping to be looked upon with respect, reverence, or devotion as a higher being.
7. Being tossed around by the ups and downs of momentary successes and failures.

I doubt that anyone who has honestly undertaken an inner journey is immune to some or all of these pitfalls. There is an enormous gap between where you find yourself today (totally dependent on the active mind) and the reality you need to unveil. It's not waking up that is painful but the pitfalls. They are created by the ego-personality, which thinks, mistakenly, that it deserves its share of the goodies that are about to fall from the sky.

If you focus on your experience here and now, the ego is irrelevant, and many distractions can be avoided. Think about parenting. Parents go through all kinds of troubles, worries, everyday crises, and arguments with their children, but without a doubt they know they love them. On the direct path you constantly reinforce your purpose, which keeps the distractions at bay. The ancient Vedas declare that everyone is defined by their deepest desires. Desire leads to thoughts, thoughts to words and actions, actions to the fulfillment of desire. So in a very basic way, desire is all you need. If your deepest desire is to wake up, to escape the illusion, to unveil reality, and in the end to know who you really are, the message gets through. Your deepest desire activates a level of awareness that will get you to the goal.

Strange as it sounds, the direct path is valid only if it leads to the place where you already are. Each of us is already a conscious, creative being

We are already in the state of wholeness, despite our allegiance to mental constructs that impose all kinds of limitations on us. The only change that's needed is a shift of identity, and yet no change is more earthshaking. Once you *realize* that you are whole, the transformation from human to metahuman has occurred.

## Nature's Hidden Message

What makes us whole already is that Nature is whole, and we are part of Nature. Such reasoning is totally sound, but it is hard for many people to accept. Modern physics holds that the universe operates as a whole, each subatomic particle woven into the cosmic fabric. But the method of science is to subdivide Nature into smaller and smaller increments, and as the investigation arrives at finer levels of existence, wholeness is lost from view. This is more than a technical glitch. A crucial question is at stake. Does the whole control the parts, or is it the reverse—there is no whole except for an accumulation of parts? A new car is a whole thing; we see it as one image and refer to it with one word. But just as easily we can see a car the way a garage mechanic does, as a collection of parts: carburetor, driveshaft, transmission, and so on. With little or no effort, your mind can shift from one viewpoint to the other.

The same holds true for how you see yourself. In the mirror you appear as a whole thing, a body; you refer to yourself by one word, your name. But if you suddenly feel acute pain in the region of your appendix, you rush to the doctor—a kind of garage mechanic for tissues and organs—and you become a collection of parts. You have the option to see yourself as whole or as a collection of parts.

When we come right down to it, what is wholeness, anyway? It's hard to imagine a more baffling question. But without an answer, living in wholeness is impossible. Fortunately, Nature presents some inescapable clues, going back to the first appearance of life on Earth. Life began 3.8 billion years ago when RNA appeared and had the ability to divide and replicate itself; a billion years later, cells with outer membranes appeared on the scene; jump ahead another billion years or more and single-celled organisms evolved into multicelled organisms.

The leap to multicelled organisms wasn't due to physical necessity—it was a creative breakthrough, and an astounding one. For almost three billion years, single-celled organisms were thriving, mutating into new species on an endless conveyor belt (which is still running at mid-ocean, where millions, perhaps billions, of undiscovered single-celled creatures ride on its surface). Life forms restricted to single cells, or even more primitive bacteria and viruses, vastly outnumber, by far more than 100 to 1, multicelled life forms. DNA had achieved the ability to fend off environmental hazards and laughed at death. The trillions of amoebas, blue-green algae, and simple fungi populating the Earth today are clones of one ancestor—it died as a physical object but is nearly immortal as a package of knowledge inside DNA.

With such a successful enterprise showing no signs of waning, there was no reason to risk multicelled life, except for one: the sheer exuberance of creativity. The basic challenge of creating multicelled life was how to reproduce a complex organism whose moving parts aren't the same. The task was like handing someone the handlebars of a racing bike and asking for the whole bike to be built from it, with the added challenge that no one had ever seen a whole bike before. In your body, two stem cells might sit in suspended animation for months or years, mixed in with fully formed cells. When a stem cell is triggered into activity, it specifically turns into a blood, brain, liver, or skin cell, not a generic human cell. The last generic cells in human development disappeared in the womb by the fifth week of pregnancy, when a ball of generic cells (the zygote) proceeded to the stage of the embryo.

The early embryo doesn't look anything like a mini–human body, but its blobby pink appearance is deceptive. As the zygote emerged from a single fertilized egg, and then went from two cells to four, eight, sixteen, and so on, DNA was biding its time. With precise timing, new instructions were sent out that altered the destiny of the embryo, pushing it closer to human form. Each new cell was given its own separate identity, through a bewilderingly complex set of chemical signals that is even more remarkable than DNA alone. We are far from being able to explain how a cell understands what to do, but we know the general outline. In the coming weeks of embryonic development not only will a brain cell become different from a liver cell, but each brain cell travels to a specific location, hooks up with similar cells that have made the same journey, and undertakes a joint enterprise: creating a brain. The same for each liver cell.

This process, known as "cell differentiation," has been minutely studied. It's fascinating to observe so-called neuronal migration as nascent brain cells, starting from their birthplace, are provided with slick pathways to slide along so that each reaches the region of the brain that needs to develop into vision, the fight-or-flight response, emotions, higher thought, and so on. (I'm giving a simplified picture, since neuronal migration is multistaged and extremely complex.) To all appearances, a cell is a squishy bag of water and soluble chemicals, but it is actually the repository of every bit of knowledge pertaining to the history of life on Earth.

Creation looks like a process unfolding step by step, bit by bit, but behind this appearance, one reality creates, governs, and controls everything. Dividing life into its component parts—the physical bits and pieces—is totally artificial. Whether we go back three billion years or to the moment you were conceived in your mother's womb, the same hidden message is present. The whole is contained in the parts. Without the whole, the parts are meaningless.

# Self-Regulation:
# The Glue of Existence

If the whole creates, governs, and controls the parts, can we observe it happening? Science doesn't deny that the universe operates as a whole, but scientists maintain their insistence that physical forces glue everything together. As with everything else in virtual reality, a story has been made up to satisfy the belief that the universe *must* be physical. The glue that really holds creation together isn't physical at all. It consists of *self-regulation*, or the ability of every system to remain intact inside its own fenced-in world.

Self-regulation is why brain cells know that they aren't supposed to be liver or heart cells. Self-regulation is why your body doesn't fly apart into a cloud of hydrogen, carbon, oxygen, and nitrogen atoms. Self-regulation has no physical properties—it isn't matter or energy—because it is a quality of wholeness itself. To show the exquisite delicacy of this invisible glue, consider how the rise in temperature of ocean water by only a few degrees in recent decades led to the destruction of coral reefs around the world. (There are other reasons as well, including water pollution, predators, and disease.) The phenomenon is known as a "marine heat wave"; such waves began to assault Australia's Great Barrier Reef in 1998 and 2002, but created only limited damage.

When water temperatures suddenly rise, a reef undergoes self-destruction through "coral bleaching"—the corals, which symbiotically depend on algae inside them to survive, become stressed and expel the algae, which also had given the reefs their bright colors. When the algae leave, the coral becomes bleached white and dies. The sudden death of huge portions of the Great Barrier Reef occurred over nine months in 2016, when a marine heat wave affected three-quarters of the world's reefs (such heat waves had begun to intensify starting in 2014). "We lost

30 percent of the corals in the nine-month period between March and November 2016," said a spokesman for the study group that monitors the reef. More marine heat waves came in 2017, affecting all parts of the Great Barrier Reef, including central sections that had managed to withstand previous destruction. The fastest-growing corals can replenish themselves in ten to fifteen years, but marine heat waves of destructive force are returning on average every six years.

This has all been driven by our failure to keep ocean temperatures from rising 2°C since preindustrial times, or roughly two hundred years. A minute change in temperature you could barely register by putting your hand in warm water was enough to tip the balance of a complex system of self-regulation that goes back 535 million years. But the lesson reaches much deeper. A coral reef is a megasystem that encloses smaller systems: all manner of fish and other marine creatures, single-celled organisms lower down on the food chain, the cells that are the basis of life, and DNA itself. Each system has its own rules of self-regulation. They build up their own boundaries for survival. Yet when viewed from the megasystem, encompassing the entire reef community, wholeness dominates over separation.

We can bring this lesson home in our own bodies. When you cuddle a baby, its warm, smooth skin feels nice, but it's also the sign of a potential threat if that warmth starts to drop off. Starting at normal body temperature, which is between 97.7°F and 99.5°F, we start to feel chilly around 97°F, and hypothermia starts to set in under 95°F, the same slight variance that threatens coral.

If a person's core body temperature drops to 90°F, a medical emergency is declared, and severe symptoms like delirium and hallucinations set in. Below 88°F, the body becomes comatose. Irregular heartbeat threatens death, which almost certainly occurs below 75°–79°F. At any given time, while healthy, we are barely 20 degrees from dying—less than the temperature swing during a summer day in the Rockies. How do we remain in check? The basic question was probably answered during the age of the dinosaurs. Current speculation is that, unlike reptiles, dinosaurs were warm-blooded.

This again was a creative leap with no pressing physical necessity. Cold-blooded creatures had existed for over a billion years as both single-cell and multicelled organisms. They breathed, ate food, expelled waste, reproduced, and survived the harshness of environmental assaults. This all happened, and continues to happen, without a key ingredient: converting some of the energy contained in food into heat, enough heat to keep the body's internal temperature constantly high enough for survival, even when the outside temperature fell too low.

Self-regulation is found at every level of Nature, beginning with the atom, which keeps itself intact without either running down like a wind-up toy or flying apart into its smaller parts (electrons, protons, and neutrons). This is enough to prove that self-regulation is how wholeness operates without needing matter or energy as its glue. Moreover, in a complex system like the human body, each cell knows that it must live for the whole, not for itself alone. Cells that opt to be selfish and multiply unchecked are cancerous, and the reward for their rampant division is death for them and the body. Normal, healthy cells do everything—eat, expel waste, reproduce, heal, and die—with the survival of the body as their main goal, not individual survival.

# The Sense of Self

This discussion about self-regulation gives us a toehold on how wholeness works. I've been insisting that everything in creation is actually a mode of consciousness. A cell is one mode of consciousness, bringing with it all the qualities of consciousness, including knowing. It is literally true that a brain cell knows it is a brain cell. Self-regulation isn't a mechanical process. It grows out of a sense of self. As a person, you have a sense of self that encompasses hundreds of self-regulating systems, just as a coral reef does. You are the wholeness that the parts need to exist. We can call this a "top-down theory," because there can be no self-regulation without the entire universe having a sense of self from the very beginning. The whole creates, controls, and governs every event.

The opposite is a bottom-up theory, the mainstream scientific view that the parts came together to assemble the whole. But self-regulation was never physically created. It is part of how consciousness operates. But we've only gotten halfway in our investigations. The discussion up to now is more than enough to prove that you are a wonder of self-regulation, as far as the body goes. But waking up is about the mind, and, as everyone knows, our thoughts can be wild to the point of madness; in any event, the next thought anyone has is totally unpredictable. You might think of us as unruly passengers, riding a perfectly assembled vehicle through life, with the body operating as a whole while the mind roams recklessly from one thought, feeling, sensation, and emotion to the next.

The direct path is based on reality being one thing, a wholeness we can live in daily life. It seems far-fetched to believe that we live with a whole mind. Everyone's mind seems like a junk pile of miscellaneous impulses tossed in a heap, from which we struggle to extract reasonable, socially acceptable behavior. This disparity between an exquisitely regulated body and a recklessly unregulated mind hasn't gone unnoticed by neuroscience and has created a baffling mystery. Thoughts, according to neuroscience, are created by brain cells. Brian cells operate by fixed laws of electromagnetism that leave no wiggle room. Electrical impulses and chemical reactions are without free will. They behave the same in a brain cell as they would in a flashlight battery or household current. How, then, did this fixed, deterministic setup lead to freedom of thought?

One answer, which seems peculiar but is widely held, notes that we don't have freedom of thought. We only think we do (which would constitute a free thought, but let it pass). The churning of brain activity is so complex, the theory goes, that we cannot find out where any particular thought comes from. But since it *must* come from brain activity, a thought is just as predetermined as the electrochemical reactions that produce it. This hypothesis neatly skips over the total unpredictability of our thoughts and feelings. If you don't have an all-seeing eye that can peer into the firing of 100 billion brain cells, of course your thoughts will seem unpredictable. Only a supercomputer could process such a mountain of information, and if computer technology continues to expand at an exponential rate, such an all-seeing eye will soon exist.

In this view, artificial intelligence (AI) is superior to the human brain, not only because it processes more information but because it is free of glitches like depression, anxiety, low IQ, wayward emotions, and forgetfulness. We must never sell technology short, or so the thinking goes. Anticipating the rise of AI to godlike status, Anthony Levandowski, known in Silicon Valley for his contribution to driverless cars and as a pioneering visionary in AI, gained media attention in 2017 by founding the first AI church, which he named the Way of the Future. Levandowski is searching for adherents, and foresees an AI godhead as not ridiculous but inevitable.

As he told an interviewer from *Wired* magazine, "It's not a god in the sense that it makes lightning or causes hurricanes. But if there is something a billion times smarter than the smartest human, what else are you going to call it?" What saves the Way of the Future from being a lampoon is the enormous impact that AI is going to have everywhere. The *Wired* interviewer writes, "Levandowski believes that a change is coming—a change that will transform every aspect of human existence, disrupting employment, leisure, religion, the economy, and possibly decide our very survival as a species."

Everyone is free to worry about a hollow, dehumanized AI future, populated by false gods, but other doomsday scenarios might befall us first. A supercomputer with global reach almost certainly will be weaponized (or will weaponize itself) into a super-hacker, capable of doing immense harm, from disabling security defenses to wrecking the banking system. Such attacks, unfortunately, are under way every day already.

Be that as it may, the *artificial* in AI keeps computers from being alive and conscious—increased speed, memory, and complexity only improve the imitation of thought. The imitation isn't thought itself. A computer will never have a sense of self, which is a primary trait of consciousness. After all, it was our sense of self that led to computers in the first place. "I can think" is part of being self-aware. Once you say to yourself, "I can think," you have a reason to build a machine that imitates thought.

The sense of self tells you that you are you. It tells you that you are alive, thinking, feeling, wishing, dreaming, and so on. No physical process created your sense of self. You possess it as a built-in trait. Imagine that you are sitting in the dark at a movie that totally captivates you. Your eyes are glued to a chase scene, perhaps; your ears are overwhelmed by gunfire, screeching tires, and police sirens. At that moment, you don't feel the weight of your body or the sensation of the seat you're occupying. You don't sense your breathing; the temperature in the theater isn't something you're likely to notice. Engrossed in the movie, you've surrendered to its spell. But with your body and your surroundings out of mind, and probably with no thoughts running through your mind, are you so captivated that *you* disappear? No.

You are present no matter how exciting a movie becomes, and the same is true of the everyday movie you walk through like a lucid dream. You can subtract everything from your experience except your sense of self. In a typical day, the experiences you notice are a tiny fraction of the sensory input you receive. You will remember a few things or nothing, but surely not everything. Most experiences flicker past and vanish, escaping all notice. But you cannot abandon the sense of self. Putting it out of mind doesn't alter, distort, or destroy it. Like a cork bobbing to the surface, it will always return.

Unexpectedly, we are plopped back onto the direct path. Your sense of self is the thing that makes you whole already. Everything that exists "out there" or "in here" is glued together by your sense of self, because it is the common denominator in every experience. Once you see this, you can identify with your sense of self, and then the direct path has reached its goal.

# Passing the "So What?" Test

As intriguing as this discussion has been, "so what?" Why should you and I go to the trouble of following the direct path? Looking at their lives, most people feel happy, more or less, with the choices they've already made. Or so they tell pollsters, who find decade after decade that more than 70 percent of responders, when asked the simple question "Are you happy?" say yes. News stories appear naming the happiest countries in the world—the latest candidate is Denmark—and correspondingly there are unhappy countries, which turn out to be those that are very poor, afflicted with war and strife, and where people must struggle to acquire the basic necessities of life.

The direct path and everything else we've been discussing—waking up, enlightenment, wholeness, the true self, metareality—confronts the "So what?" test. We make choices based on whether they matter enough in our lives. This isn't a trustworthy measure of what is actually beneficial, however. Cosmetics and fashion matter enough to millions of people, just as fantasy football and adding more guns to their collection matter enough to millions of others. "So what?" is highly personal and unpredictable, but it is also ruthless. Until something matters enough, we won't change our accustomed way of life.

Here's my answer to "So what?" as it applies to the direct path. In a very real way, becoming metahuman is about eliminating everything that is unnecessary for our species. Such a process is already under way by bits and pieces. The modern world has eradicated—or is in the process of eradicating—many things that *Homo sapiens* can do without. It would have seemed impossible once to do without religion, judging by how a belief in God or the gods lay at the heart of every ancient culture. But there are millions of people who have adopted a secular worldview, placed their faith in science, and live without a sense of loss in the absence of religion.

We can argue forever over whether religion is good or bad for society, and over every other thing that is in flux. From the viewpoint of metahuman, however, stripping away all the things that *Homo sapiens* doesn't need anymore is essential. The illusion is getting thinner, so to speak. Mental constructs like war and poverty are no longer considered facts of life humanity must accept. Why not get rid of the whole illusion, then? Why not do it simply, completely, and with the least discomfort possible?

Once you eliminate everything that isn't real, what's left must be real. The direct path accords with this notion. It holds that only one thing is real if everything unreal is taken away: the sense of self. Every life has its own priorities. One person's suffering may be rooted in poverty, while another person's suffering is rooted in sickness, a bad relationship, or lonely old age.

Nothing is more tangled and chaotic than virtual reality. The process of dismantling every cause of suffering is too complex to figure out in advance. The only realistic answer is to allow the illusion to dissolve as effortlessly as it developed in the first place. This is what the direct path is all about; this is what sets it apart from every other scheme for the improvement of human existence.

It seems too good to be true that mind-forged manacles will simply drop away of their own accord. But that's exactly what Rumi meant when he declared that the jail cell is open. Since we are already free, there is no effort needed to get free. You are free once you base your life on one thing, your sense of self.

But as things stand, the sense of self is almost totally overlooked. Woody Allen's 1983 movie, *Zelig*, depicted its title character as a kind of historical phantom. Even though he was totally insignificant, Leonard Zelig was a human chameleon. He had the ability to melt into his surroundings completely. The movie takes place in the 1920s and 1930s, and the novelist F. Scott Fitzgerald is the first to notice Zelig's amazing transformations. At a *Great Gatsby*–like party, Zelig is in the living room talking in a refined Boston accent and upholding Republican values, while later in the kitchen he comes off as a common man with Democratic ideas.

Somehow Zelig is everybody and nobody, everywhere and nowhere. That's a parable for the sense of self. It is present in every experience but melts invisibly into the background—until you start paying attention to it. Then the sense of self, unlikely as this seems, takes center stage.

The first step is to begin to notice your sense of self. Imagine a debating society where one side is defending a controversial position on the Middle East conflict, abortion rights, or racism. See yourself standing up and delivering an eloquent argument on one side. Let's say you vehemently oppose racism, believe in a two-state solution for the Palestinians, and support a woman's right to have an abortion.

Now put yourself on the opposing side and make exactly the contrary argument for each issue. It might help if the contrary position is outrageous—you might see yourself defending the necessity to ban all abortions or supporting the terrorists who deny the right of Israel to exist. Even though you will experience resistance in abandoning the side you actually believe in, debaters do it all the time. They shift perspective from pro to con at the drop of a hat. This ability to shift perspectives is beyond any set of beliefs. We can put on any coat we want to wear. Yet whether you speak up for ideas you love or ideas you hate, *you* are always present. Your sense of self is independent of anything you think or say.

Next example: Take any common object in your house and touch it. Is it you who touched it? Anyone will automatically say yes. Look at the same object. Is it you who sees it? Yes again. There is no time in the history of humankind when the answer would be different. What keeps changing is our mental model of what it means to touch and see an object. In prehistory, before language arose, there was no explanation for seeing and touching. In an age of faith, when dead flesh, the body, was animated by the soul, it was God who made sensation possible. Today, the experience of seeing and touching is ascribed to the central nervous system and the activity of the brain in receiving sensory data from the world "out there."

Yet if you don't bother with any of these explanations, there is only the self having an experience. You can modify this exercise in other ways. In medical school, students are taught that bodily sensations reach the brain through a spidery web of afferent nerves running everywhere. If you touch your hand or lift your arm, the signals from afferent nerves are responsible for feeling the experience. Every signal is like an Inca runner dashing from a remote region to the Andes to the emperor in the royal city of Cuzco. The information from afferent nerves is constantly flowing, so why don't we feel every limb's weight, position, warmth, and the like all the time? Sit in your chair and let your attention roam from the top of your head to your nose, then to your heart, then to the tips of your toes.

If I ask, "Who is roaming from one sensation to another?" you will surely answer, "I am." The self selects whatever it wants to pay attention to. Therefore, the self is not bound by the activity of the nervous system. It is present no matter which nerve cells are firing.

One final example: If you close your eyes and see the color blue, this experience is happening in your awareness. If you open your eyes and gaze at the blue sky, where is the blueness? It is still in your awareness. The experience of blue "in here" and blue "out there" occurs in the same place. When you handle an object or smell it or hear it, you can remember those sensations. Did the sensations travel from the outside world to your inside world? We assume they did. Sitting around a warm campfire and watching its flickering flames is different from remembering the warmth and the flickering flames. But both experiences happen in awareness. Hearing a bang in your head and hearing a car backfire on the street are two different experiences, but they share the undeniable fact that both occur in awareness.

If awareness is present both inside and outside, it is independent of space. Likewise, the campfire you sat beside might have been in childhood, while the memory of it is now. Since awareness was registering the campfire both then and now, it is independent of time. Once you see that awareness isn't bound by time and space, then *you—the real you that is pure consciousness*—must not be bound by time and space. Wherever your awareness goes, your sense of self is present; the two are fused. The self is present in everything, but we don't notice this because, like Zelig, the self knows how to melt perfectly into every situation.

I think most people, with a little reflection, can accept that they have a sense of self following them everywhere like an invisible shadow. But to fully pass the "So what?" test, the sense of self can't simply be a bystander. Part of Woody Allen's conceit was that, if you look closely, Zelig appears in photos of momentous events like a presidential inauguration—but this didn't make him important. Similarly, sensing yourself in every experience doesn't put it center stage. The direct path brings the sense of self to light. What's needed next is to discover the enormous difference this modest attainment actually makes.

The leap from human to metahuman, the shift from virtual reality to the "real" reality, is all contained in the simple act of noticing the self. Not the ego, which wants to hog the limelight and claim that it and it alone is the self. The ego-personality can't help but contain the sense of self—every experience contains it. But the ego blocks our view. It is tied to virtual reality and keeps luring our attention everywhere *but* the sense of self.

As we'll see in Part Three, once you rid yourself of all the things you don't actually need, your ego can be discarded, too, but not your sense of self. As the unnecessary burdens and rubbish of the spell/dream/illusion are tossed out, the sense of self doesn't hog the limelight. Being the only thing about you that is timeless and eternal, it doesn't have to do anything. It simply is, a beacon of pure awareness that shines constantly in every second of your existence. Ironically, everything else in your life has a much harder time passing the "So what?" test.

PART THREE

BEING METAHUMAN

# 10

FREEING YOUR BODY

As we saw, consciousness is everywhere, always, and everything. If that's true, then you are everywhere, always, and everything. But in your daily life, even if you were totally convinced about this, you are guided in the opposite direction. You've endured a lifetime of training and conditioning telling you that you are a solitary person sitting by yourself in a room. Instead of always, your life span is very limited, bookended by two events—birth and death. Instead of everything, you are a bundle of very specific things, beginning with your name, gender, marital status, work, and so on.

The virtual reality we want to dismantle is made up of many moving parts that occupy their own compartments. One part is the body, another the mind, the world, and other people. These parts were set up so that life can be managed one piece at a time. You go to college for your mind, to the gym for your body, on a date to establish a relationship, to work to bring home money.

From a metahuman standpoint, *any* division of life into bits and pieces only supports the illusion. Wholeness is wholeness, not a collection of parts. To put it another way, life happens all at once, here and now. The reason we cling to virtual reality is that the prospect of "all at once, here and now" is too overwhelming.

After pondering how to bring this truth home in a way that's practical and relatable, I concluded that the direct path should start with the body, not the mind. My reasoning is that the body is what holds most people back. They experience themselves encased in a body. They accept the reality of birth, sickness, aging, and death. They seek physical pleasure and shrink from physical pain. As long as these are the preconditions of your daily life, you can't be metahuman. Your body won't accept it. A liver, heart, or skin cell can't scream, "Are you crazy?" But it does seem crazy to abandon the body as a physical object—this act of craziness makes your body the perfect place to start. If you can get past the body as a package of flesh and bone, transforming it into a mode of consciousness, everything else naturally falls into place.

## The Anatomy of Awareness

If we have trapped ourselves inside an illusion, the body must be part and parcel of that—and it is. Your body is your story in physical form. As your story grew over the years, the things you thought, said, and did required a huge array of brain activity. Learning to walk was a triumph of balance, eyesight, and motor coordination, but once you as a little child put together this complex puzzle, your brain remembered everything you learned and stored it for life, so you can move on to something new. Your brain has stored a host of skills, from speaking and writing to riding a bike, doing arithmetic, and dancing the waltz. These mental attainments are embodied in you physically.

But as soon as this is pointed out, we risk falling back into dividing "mental" from "physical," which sends us straight back into the illusion. As babies learn to walk, they are under no such illusion. As they totter, fall, get up again, and keep on trying, the experience fits the holistic description given above: the whole thing happens all at once, here and now. The same is true of any skill you can name—learning it wasn't mental and/or physical. It was happening in only one dimension: awareness.

There are so many ways that we put mind and body in opposition that going into all of them would be impossible. The direct path doesn't even require the small amount of discussion we've just passed through. Instead, it experiences the body *in awareness*. Once you do that, the division between mental and physical returns to the authenticity of a baby learning to walk. You return to the self as the agent of the whole experience, fusing thought and action. In this case, any action is returned to wholeness, which is where it occurs.

To begin, just be open to the idea that your body isn't a physical object in which you reside. Such a viewpoint is just a habit of thinking, even though a stubborn one. I will guide you through an exercise that will give you the direct experience of living, not in a physical body, but in awareness. (This exercise and the ones that follow are much easier to do if someone reads them aloud to you. If you can find a partner who will join you, switching the roles of reader and participant, all the better.)

# Exercise: The Body in Awareness

This first exercise involves the following steps:

STEP 1:  Being aware of your body.

STEP 2:  Being aware of some bodily processes.

STEP 3:  Being aware of the body as inner space.

STEP 4:  Expanding inner space beyond the skin.

STEP 5:  Resting in wholeness.

Each step is a natural progression from the one before, and each step is a simple, direct experience. You don't have to memorize the instructions, just go through the exercise as an experience.

## STEP 1: BEING AWARE OF YOUR BODY.

Sit quietly with your eyes closed. Let bodily sensations be your focus of attention, so that you can feel your body. There's no need to try to stop your thoughts; they are irrelevant. It doesn't matter what sensations you happen to feel. Just be with your body.

## STEP 2: BEING AWARE OF SOME BODILY PROCESSES.

Be aware of your breath as it goes in and out. Slow your breath down a bit, then make it go a bit faster. Move your attention to the center of your chest and be aware of your heartbeat. Take some easy, deep breaths and sense your heartbeat slowing down as you relax. See if you can feel your pulse elsewhere—many people can feel it in their fingertips or forehead or inside their ears, for example.

## STEP 3: BEING AWARE OF THE BODY AS INNER SPACE.

Now move your attention to the inside of your body. Feel your head as empty space, and start going down your body to the chest, stomach, abdomen, legs, and feet, pausing at each place to experience your internal organs as a space in which your awareness freely moves. If you want, you can also experience your breath as the expansion and relaxation of the space in your chest, your heartbeat as a constant pulsation in the space of the chest.

## STEP 4: EXPANDING INNER SPACE BEYOND THE SKIN.

Once you have felt the inside of your body as empty space in which bodily processes are taking place, run your attention across your skin. Roam over the sensations in your head, feeling the outlines of face, scalp, and ears. Move downward, letting your awareness go to every other place— throat, arms, hands, feet—as sensations reach you.

Now lift your awareness slightly above your skin and gently let it expand beyond the contours of your body. Some people can do this easily, but others need to conjure up an image—you can see your inner space suffused with light and watch the light expand until it fills the room. Or you can visualize inner space around your heart as a sphere, a round balloon that expands a bit more every time you inhale, watching it get bigger and bigger until it takes up the entire room.

## STEP 5: RESTING IN WHOLENESS.

Once you have gone through the previous steps, rest quietly for a minute or two. Let the experience of the body be here now.

What this exercise has done is multifold. You have freed yourself from being trapped inside a thing, substituting the experience of your body as processes and sensations. Since your experience was conscious, these processes and sensations moved to where they actually occur, in your awareness. Then you opened yourself to experience your body as inner space—in Sanskrit this space is called *Chit Akash*, or mind space. Since everything in life happens in mind space, you expanded the space until there was no boundary between "in here" and "out there."

You may be surprised that you've accomplished all these things, and since the experience was likely to be unusual, you will easily pop back into the habit of feeling that your body is a physical object you reside in, like a mouse hiding in the walls of a house or a rabbit in its burrow. Yet, at the very least, you now can see that there is an alternative to the old habit. To free yourself from your body, the old habit won't serve you.

To break down the old habit—a process I call "thawing," as discussed before—take time once or twice a day to repeat this exercise. Once you become used to it, the whole thing runs naturally and smoothly. We're all practical-minded, so what good is the exercise in daily life?

- You can do it when you feel stressed.
- You can do it to relieve tightness in your body or other unpleasant and painful sensations.
- It also works to help relieve worried, anxious thoughts.
- You can center yourself with this exercise whenever you feel distracted and scattered.

Various practices in yoga and Zen Buddhism bring the body under control through sheer focused awareness. When you felt your heartbeat and breathing, for example, you took the first step to controlling both processes at once through the vagus nerve, one of the ten cranial nerves that extend from the brain to the rest of the central nervous system.

The vagus nerve is a wanderer, tracing its course like the trunk line of an old-fashioned telephone system from brain to neck, down into the chest, past the heart, and extending into the abdomen. It is the longest cranial nerve, and its fibers both send and receive sensory information. Most of the sensations you experienced in the exercise you just did were channeled through the vagus nerve.

You can take advantage of this knowledge in a practical way through "vagal breathing," which consists of inhaling to the count of four, holding the breath to the count of two, and exhaling to the count of four. This simple rhythm of 4-2-4 is easy for a normal healthy adult, and no one should try to force it to the point of gasping or feeling uncomfortable. To the great surprise of medical researchers, vagal breathing is the best way to reduce stress, particularly the immediate signs of ragged breath, increased heart rate, and muscle tension.

It turns out that chronic, low-level stress is more important and more pervasive than acute stress. In modern life, being in a state of chronic low-level stress is so common as to be accepted as normal. But your body doesn't experience it as normal at all; the earliest beginnings of heart disease, hypertension, sleep and digestive disorders, and probably some cancers can be traced to chronic stress.

It is no accident that the location of these disorders parallels the course of the vagus nerve, which serves to communicate stress to the heart, stomach, digestive tract, and then the rest of the body as the nervous system branches out. Vagal breathing brings the state of the body back into balance and relieves tension. So even if you haven't come to the realization that everything physical and mental happens in consciousness—meaning that our body happens in consciousness—here is an unmistakable clue.

We accept that a skill like walking and riding a bike is permanent once it has been learned. But, at a more basic level, the biorhythms that sustain the bodymind as a whole were learned and absorbed millions of years ago by our hominid ancestors. Modern life pushes us to unlearn them, as witnessed by the huge number of people with digestive and sleep disorders. Both of those functions are controlled through built-in biorhythms. Once your body has forgotten how to express a biorhythm, the effect is like one trumpet or violin playing a different piece of music from the rest of the orchestra—the whole symphony is ruined.

Vagal breathing may wind up being very useful beyond its ability to normalize heart rate, lower blood pressure, and make breathing more regular. It's a good practice to use in bed just before you fall asleep. Mild to moderate insomnia can often be eased or totally healed. General stress is eased, including stressful thoughts and a racing mind, which countless people experience when they try to go to sleep.

I've gone into a little detail (the whole topic is covered in depth in *The Healing Self*, which I coauthored with Rudy Tanzi) to make the point that a physical map of the nervous system, the operation of the vagus nerve, consciously intervening in its operation, and sensing the results directly are *all one thing*. The direct path leads us back to that one thing, which will, in time, completely end the state of separation and allow us to rest in wholeness, which is reality.

# Your Body, Your Story

Once you begin to experience your body in awareness, transformation starts to occur. You are moving out of the separation between mind and body toward wholeness. You need to be whole to be who you truly are. The direct path is experiential. It's not theoretical, nor is it therapy or a spiritual journey.

As you experience your body in awareness, you have taken time out from your story. This seems like a modest thing, a moment or two spent in doing nothing but being here. Yet there's no other way to dismantle your story that is this direct. Millions of people benefit from therapy and walking the spiritual path. But eventually we must all face the fact that *everything* we do, even in the name of healing and spirituality, is taking place inside our story. As a result, we can get glimpses of wholeness— these are often quite beautiful and uplifting—yet they do not carry us to metareality as our home.

At the moment you are your story—it can't be helped—and this keeps you trapped in separation. The word *separation* may not come to mind; you might not even consider it a problem. To illustrate, consider the following sentences, which all of us have said or heard:

I hate my body.
You're only as old as you think you are.
Youth is wasted on the young.
I used to have a perfect figure.

The statements express different sentiments, but each reflects the separation of mind and body. "I hate my body" comes from someone who feels trapped in physicality. The person is bemoaning what she has done to her body, or what her body has done to her. The state of separation is obvious. "I" is playing the part of the victim, and "body" is the culprit.

"You are only as old as you think you are" is much more optimistic, asserting that the mind can overcome the deterioration of aging. Yet, as we all know, this is in part wishful thinking. Aging is an inexorable process. It is much better to have a positive attitude toward it than a negative one—society is benefiting from the "new old age," which envisions every stage of life as vigorous, productive, and healthy. But a thought or a belief isn't the same as a state of awareness. "You're only as old as you think you are" cannot substitute for the true self. When established in your sense of self as a permanent state, aging holds no threat because you identify with the timeless (we will go into what the *timeless state* means in the next section). A good attitude toward the aging process still leaves you trapped in your story.

Oscar Wilde's quip "Youth is wasted on the young" puts a witty face on a sad wish that many people have: If only I could go back in time, I'd live my life so much better. Regret over the past is mingled into everyone's story, and the basis for this regret (and its opposite, which is nostalgia) is that the passage of time has power over us. "I used to have a perfect figure" states this more directly by connecting the passing of the years with a loss of physical attractiveness, while implying that "a perfect figure" is the same as self-worth and sexual desirability.

These examples of how the mind feels separate and different from the body could be expanded endlessly. The body is subject to all manner of judgment, and yet the underlying process, whether you love your body or hate it, hasn't been examined closely. This underlying process is the inescapable way your body has absorbed every detail of your life story and now mirrors it. To be trapped in your story and trapped in your body are the same thing. Your brain has been shaped every minute since you were born, and it has communicated everything you experience to the body's fifty trillion cells, which in turn pass the messages on to your DNA.

Wholeness, then, isn't just a mental shift. It's a revolution in the body-mind that undoes the past, beginning in the brain but reaching out to liberate every cell and influencing the genetic activity in each cell. To see how this revolution occurs, let me go to the heart of every story, which is time.

# How to Be Timeless

There was a period in infancy when your experience was original and authentic. You were too young to interpret the world on your own. All your development was occupied with the basics of walking, talking, exploring the world, and so on. Let's call this the prestory period of life. William Blake divided one group of poems into "songs of innocence" and "songs of experience," which were a nonbiblical narrative of the Fall from Grace. Like the romantics who were to follow, and who idolized him, Blake believed that the Fall didn't occur to Adam and Eve; rather, it happened to children as they lost their innocence. The Fall was a repeated experience generation after generation.

What Blake saw in innocence was a fresh, simple, lyrical, and joyous perspective on the world. The tone is set in one of the most famous songs of innocence, "The Lamb," where you can read *infant* for *lamb*:

> *Little Lamb who made thee*
> *Dost thou know who made thee*
> *Gave thee life & bid thee feed,*
> *By the stream & o'er the mead;*
> *Gave thee clothing of delight,*
> *Softest clothing wooly bright;*

*Gave thee such a tender voice,*
*Making all the vales rejoice.*

In contrast to this vision of a childhood Eden, the songs of experience are bitter and dark, reflecting the hardship Blake knew firsthand and saw all around him in eighteenth-century London. One famous poem, "A Poison Tree," reimagines the tale of original sin as the dark side of human nature, in verse that could be a nursery rhyme:

*I was angry with my friend:*
*I told my wrath, my wrath did end.*
*I was angry with my foe.*
*I told it not, my wrath did grow.*

*And I watered it in fears,*
*Night & morning with my tears:*
*And I sunned it with smiles,*
*And with soft deceitful wiles.*

*And it grew both day and night,*
*Till it bore an apple bright.*
*And my foe beheld it shine,*
*And he knew that it was mine.*

*And into my garden stole,*
*When the night had veiled the pole;*
*In the morning glad I see,*
*My foe outstretched beneath the tree.*

Even without the benefit of Blake's vision, we have all experienced the transformation from innocence to experience. All it took was time.

Nothing else was necessary as we learned the standard interpretation of the world everyone around us accepted. A baby is never bored. He looks on the world with wonder. The hours don't hang heavy; deadlines don't make a baby rush through his days. He has no hunger for distraction so that he can escape from himself.

Being a visionary, Blake saw the possibility of liberation from the fallen state, which he called "organized innocence." It's a brilliant phrase, because it implies that a person's experience can be as original, authentic, and untainted as a baby's while retaining the organized mind, a mind we must have to perform higher functions as adults (which includes taking care of babies). Returning to innocence means embracing values like love and creativity, which become more valued as we reach maturity. Yet the years of maturity make it more and more difficult to journey back to innocence. For all the countless people who yearn for what it felt like to first fall in love, few find a way back.

The culprit isn't experience, because experience can still be joyful and authentic at any time of life. The culprit is hidden in the texture of our lives—it is time. I said above that becoming embedded in the interpreted world (i.e., the spell/dream/illusion) only needed time, nothing else. By the same token, escaping the grip of time is the only way out. Far from being a mystical notion, you can be timeless right this minute—in fact, that's the only way.

For most people, the two words *timeless* and *eternal* seem roughly the same. For religious believers in the Christian and Muslim traditions, Heaven is eternal, a place where time goes on forever. For the nonreligious, time ends with physical death. In both cases, however, ordinary clock time has ceased. But there are problems with all these concepts, and if we go deeply into the subject, time is very different from what we casually accept.

Physics has had a lot to say about time, thanks to Einstein's revolutionary concept that time isn't constant but varies according to the situation at hand. Traveling near the speed of light or drawing near the massive gravitational pull of a black hole will have a drastic impact on how time passes. But let's set relativity aside for a moment to consider how time works in human terms, here and now. Each of us normally experiences three states of time: time ticking on the clock when we are awake, time as part of the illusion of having a dream, and the absence of time when we're asleep but not dreaming. This tells us that time is tied to our state of consciousness.

We take it for granted that one kind of time—the one measured by clocks—is real time, but that's not true. All three relationships with time—waking, dreaming, and sleeping—are knowable only as personal experiences. Time doesn't exist outside human awareness. There is no absolute clock time "out there" in the universe. Many cosmologists would argue that time, as we know it in our waking state, entered the universe only at the big bang. What came before the big bang is probably inconceivable, because "before the big bang" has no meaning if time were born at the instant the cosmos was born. If you go to the finest level of Nature, to the vacuum state from which the quantum field emerged, the qualities of everyday existence, such as sight, sound, taste, touch, and smell, no longer exist, and there is also a vanishing point where three-dimensionality vanishes, along with time itself.

What lies beyond the quantum horizon is purely a matter of conjecture. The precreated state of the universe can be modeled almost any way you choose, as being multidimensional, infinitely dimensional, or nondimensional. So it must be accepted that time came out of the timeless and not just at the big bang. Everything in the physical universe winks in and out of existence at a rapid rate of excitation here and now. The timeless is with us at every second of our lives.

Yet something looks fishy in that sentence, because the timeless can't be measured using a clock, so it makes no sense to say that the timeless is with us "at every second." Instead, the timeless is with us, period. This world is timeless. There is no need to wait for death or Heaven to prove that eternity is real.

Once you acknowledge that the timeless is with us, a question naturally arises: How is the timeless related to clock time? The answer is that the two aren't related. The timeless is absolute, and since it can't be measured by clocks, it has no relative existence. How strange. The timeless is with us, yet we can't relate to it. So of what good is the timeless?

To answer this question, we have to back up a bit. Clock time has no privileged position in reality. There is no reason why it should be elevated above dream time or the absence of time in dreamless sleep. Clock time is just a quality of daily life, like other qualities we know as colors, tastes, smells, and so on. Without human beings to experience these qualities, they don't exist. Photons, the particles of light, have no brightness without our perception of brightness; photons are invisible and colorless. Likewise, time is an artifact of human experience. Outside our perception, we cannot know anything about time. This seems to contradict the cornerstone of science, which holds that "of course" there was a physical universe before human life evolved on Earth, which means that "of course" there was time as well, billions of years of time.

Here we come to a fork in the road, because either you accept that time, as registered by the human brain, is real on its own or you argue that, being dependent on the human brain, time is created in consciousness. The second position is by far the stronger one, even though fewer people believe it. In our awareness we constantly convert the timeless into the experience of time—there is no getting around this. Since such a transformation cannot happen "in" time, something else must be going on. To get a handle on this "something else," let's look at the present moment, the now, the immediate present.

All experience happens in the now. Even to remember the past or anticipate the future is a present-moment event. Brain cells, which physically process the conversion of the timeless into time, function only in the present. They have no other choice, since the electrical signals and chemical reactions that run brain cells occur only here and now. If the present moment is the only real time we can know in the waking state, why is it so elusive? You can use a clock as fine-tuned as an atomic clock to predict when the next second, millisecond, or trillionth of a second will arrive, but that's not the same as predicting the now. The present moment, as an experience, is totally unpredictable. If it could be predicted, you'd know your next thought in advance, which is impossible.

As we touched on already, the present moment is always elusive because the instant you register it as either a sensation, an image, a feeling, or a thought, it's gone. So let's boil these insights down. The now, the place where we all live, can be described as:

- the junction point where the timeless is converted into time
- the only "real" time we know in the waking state
- a totally unpredictable phenomenon
- a totally elusive phenomenon

Now, if all these characteristics are being correctly described, it turns out that we have been fooling ourselves to believe that time is a simple matter of tick-tock on the clock. In some mysterious way, each of us occupies a timeless domain, and to produce a four-dimensional world for the purpose of living in it, we mentally construct it. That is, we create the world in consciousness first and foremost. Reality, including ordinary clock time, is constructed in consciousness as well.

We must not fall into the trap of saying that the mind creates reality. The mind is a vehicle for active thinking experience, and, like time and space, it has to have a source beyond something as transient and elusive as thoughts. If we trusted our minds, we'd equate going to sleep with death. In sleep the conscious mind gives up the world of solid physical objects and clock time. Yet when we wake up in the morning, there is a return of solid objects and clock time. They were held in waiting, so to speak, by consciousness, even during the eight hours a day that the thinking mind was out of commission.

If the direct path aims to take us beyond the illusion we accept as real, it must deliver the experience of being timeless. In its fullest state, the timeless experience is simple, natural, and effortless—it is the sense of self, which quietly lies inside every other experience. We are only partially there. The exercise in this chapter, which allows you to experience your body in awareness, thaws out one aspect of the illusion. The present discussion of being timeless thaws out another aspect. In both cases, as physicality and clock time start to lose their grip, you realize that there is another way to live: being metahuman.

When you start being metahuman, you exist at the timeless source of the self. It's a huge step to know that the timeless is with us, beyond any belief in birth and death, age and decay. Things appear and disappear in our dreams when we're sleeping. Yet we don't mourn them, because we know that dreams are an illusion. What matters isn't the things that appear and disappear, only that you don't mistake the dream for reality.

Discovering that the same is true about our waking dream sets us free from the fear of death. Metahuman is about more than this. Liberated from illusion, we can be free of all fear. In the end, waking up leads to absolute freedom. We don't just lead our everyday lives. We navigate the field of infinite possibilities.

# 11

# RECOVERING THE
# WHOLE MIND

I magine that you've gone to a cabin in the woods that sits by a small lake. The sun wakes you up very early, and when you walk down to the lake in the pale light of dawn, it is perfectly still. The surface is mirror-like and undisturbed. You have the urge to pick up a pebble and toss it into the lake. The pebble lands with a plop, ripples spread out across the water, and, as the ring of ripples widens, they fade away until nothing remains but the still surface of the lake. It has returned to what it was before you came along to disturb it.

In this modest experience is reflected the whole story of the human mind. All that's needed is to see the scene in reverse, like running a film backward. The surface of the lake is perfectly still. A faint ring of ripples disturbs the water, at first in a way that's barely discernible. The ripples grow bigger and the ring starts to shrink. All at once a pebble pops out of the water and flies into your hand.

This is how the human mind looks from the perspective of meta-reality. There is pure consciousness, still and undisturbed. Faint activity begins to stir, so faint that you have to be quiet and very alert to notice it. But this activity—call it a vibration in consciousness—becomes more urgent until something fully formed pops into existence: a sensation, image, feeling, or thought.

In daily life a myriad of mental events pop into existence, and they are so constant and compelling we never experience the still surface of the lake, which is pure consciousness. The direct path aims to recover this experience, because a still, undisturbed state of awareness is *whole mind*. Whole mind contains the infinite potential that is unique to *Homo sapiens*. As with so many things in metareality, whole mind is already here and now. You have to have a lake before you can throw a pebble into it, and you have to have a lake for the pebble to pop out of it, too. Both processes occur in the cosmos as every subatomic particle emerges from the quantum vacuum and fades back into it. Likewise, your thoughts arise from silence, poise in midair only as long as it takes to be noticed, and then fade back where they came from.

If whole mind is already here and now, why do we need to recover it? The answer offered in this book is that metareality lies beyond the illusions of virtual reality, which bamboozles us by seeming to be real. Whole mind is indisputably real, as the direct path aims to prove. But I feel that something more is needed. If mental activity were exactly like pebbles popping out of a lake, each thought would naturally die away, and the mind would effortlessly return to its undisturbed state. In other words, there would be no interruptions, no clinging or blurring or storms to whip the lake up.

Unfortunately, our minds are not smoothly connected to pure consciousness. Our minds are a confused, chaotic jumble, vividly described in some Eastern traditions as a monkey. Every thought joins the storm of other thoughts without returning to a still, undisturbed state. To be human is to ride the storm. Once you learned as a small child to speak and put thoughts into words, your mind was off and running. Fueled by the agenda of the ego, "I" started the endless task of accepting A and rejecting B. Even two- or three-year-olds want what they want: toys to call "mine," attention from their mother, their favorite foods, stories they want to hear over and over again. Rejection is a vital part of the ego's agenda as well. A two- or three-year-old throws some toys on the floor, rejects hugs and kisses, refuses to eat certain foods, and so on.

There is a famous cartoon by Carl Rose and E. B. White showing a mother and child at the dinner table. The caption goes:

MOTHER: "It's broccoli, dear."
DAUGHTER: "I say it's spinach and I say the hell with it."

If kids spoke in adult voices, that's exactly how they'd talk!

From the perspective of metahuman, none of the ego's agenda is a necessity. Why go to the trouble of constructing a self when you already *are* the self? The only real necessity is to be here now and allow life to unfold.

But because our thoughts don't fade away into pure, quiet calmness, an obstacle arises. Thoughts interfere with our ability to be here now. Every thought generates a new thought, every sensation a new sensation. This endless mental activity is so distracting that we cannot see our source or touch it or sense its presence. Getting rid of mental interference has become hard work. The years spent in Zen Buddhist monasteries or sitting in lotus pose in a Himalayan cave are images in popular culture that represent just how arduous it is to quiet the mind. The direct path cuts through the hard work in a simple way, as illustrated in the following exercise.

# Exercise: Eyes Open, No Thoughts

This exercise is reserved for the moment you wake up in the morning. When you open your eyes and before you get out of bed, your mind begins its day. You are preset to jump into your customary routine. You start to think about what you have to do that day, and very soon you have rebooted your personal story—your mind is trained to run automatically, like a computer whose software is ready for instructions the minute the machine is turned on.

Yet there is a small interval before your story reboots. For a few seconds you are awake but not yet engaged with the world. At this moment there is no intervening thought to block pure awareness. You can expand these few seconds into a conscious experience. Here's how.

Go to bed with the intention that you are going to take the following steps the instant you wake up the next morning:

*Step 1:* When you first are aware of waking up, open your eyes.

*Step 2:* Stare at the ceiling without focusing on anything in particular.

*Step 3:* Try to keep your eyes open. Focus on this and nothing else.

What this exercise does is to set a trap for undisturbed awareness. Each step is important in the sequence. As you go to bed the night before, you set an intention; this puts the exercise at the top of your mental agenda in the morning. Opening your eyes and staring at the ceiling distracts you from starting to think. Focusing your attention on keeping your eyes open is the real secret, however. Your brain wakes up in the morning in waves, alternating sleep and wakefulness. As these waves increase, there is steadily more wakefulness than sleepiness, and thus you wake up, still feeling a bit sleepy.

By focusing your attention on keeping your eyes open, your brain has no choice but to block thinking. The effort to stay awake occupies it entirely. (You can accomplish the same thing by visually seeing a blue dot in your mind's eye and keeping your attention on it.)

Once the steps of this exercise are mastered, you will be in a state of "eyes open, no thoughts." As you lie there, simply notice what this state feels like. Very few people have actually been awake without thinking for more than a few seconds. With practice, you can sustain this exercise for up to a minute. It's a significant thing to be still, quiet, calm, and alert with no ego to defend or story to construct. The more you are aware of how significant the experience actually is, the more insight you will have into your true self. The exercise isn't an end unto itself, merely a starting point.

There are variants of getting to the same state of "eyes open, no thoughts." In meditation, for example, the mind settles down into a rested quiet state, and when you open your eyes at the end of the meditation, this mental relaxation persists. Here the mind is coordinating a brain response, the natural tendency to return to pure consciousness if given a chance. The purpose of a mantra, if you practice mantra meditation, is to give the mind a sound without meaning to occupy it, which helps you stop focusing on what your thoughts are trying to tell you. As the mantra returns to the mind, it becomes subtler and fainter, allowing an easy slide into quiet mind. At first, when you are coming out of meditation, "eyes open, no thoughts" is brief, because we are so used to jumping back into our story. But over time there is an expansion, and "eyes open, no thoughts" turns into a steady state around the clock.

At that point, a person is considered fully awake. In the morning exercise just presented, "eyes open, no thoughts" is blank. But the experience of being fully awake isn't. Mental activity arises—pebbles pop out of the lake—while the undisturbed state remains. Silence and activity are experienced as one thing: awareness moving within itself. The experience of "eyes open, no thoughts" is modest at first, yet it opens the way to transform how we use our minds.

I've made the point that most people don't abandon their stories, no matter how much pain and suffering those stories are causing, because their natural instinct is to cling to those stories. If you aren't a battered wife who refuses to leave her husband, a closeted gay man afraid to come out, an opioid addict unable to end her habit—or any other person stuck in deep pain—you can't grasp how horrible the consequences of clinging can be. Stories can turn into a hopeless state, and that state challenges the best that medicine and psychotherapy can do to help.

Since we all cling to our stories, can the direct path really offer a way out? Let's agree that you and I and everyone else are stuck in our story. *Stuck* is convenient shorthand for old conditioning that keeps us from creating the changes we want in our lives. It's possible to learn a lot from the various ways to get unstuck that have failed. Consider the New Testament, where Jesus of Nazareth appears as one of the most compelling spiritual figures in recorded history. The story of the Son of God who preaches peace to the world and is crucified for it constitutes a drama that has transfixed the West for two millennia.

It is striking, however, that what the New Testament teaches is often impossible to achieve. "Love thy neighbor as thyself" clearly runs counter to how human nature operates (and would have been a fatal teaching to follow in the face of Nazism, at a time when many Jews had neighbors who joined the Nazi Party, or in Bosnia where mixed Muslim and Christian neighborhoods became battlegrounds based on religion). "Turn the other cheek" seems masochistic, except to a deeply confirmed Christian, asking for bullies to victimize the helpless even more than they already do. But this isn't a criticism aimed at one faith—all spiritual traditions have foundered on the same obstacle. Human nature is in love with its own self-created drama, and the more stark the contrast between good and evil, the more we cling to our story. If the human race is waiting for the day when either good or evil is triumphant, the only predictable outcome is that the drama is designed to last an eternity.

Most people are clinging to drama at a much milder level, naturally. Except for those times when unconscious forces erupt and a whole nation, or the world at large, is dragged to the edge of catastrophe—sometimes plunging over into the abyss—daily life is stuck in mundane wants and desires, duties and demands. Like being drowned in the bathtub or being drowned at sea, the result is the same.

Yet, someone doesn't have to experience misery to experience being in separation. Your quality of life may be good or bad—a life generally brings both—the situation comes down to the same nitty-gritty: stuck is stuck. The drama is self-perpetuating. Even our desire to escape the spell/dream/illusion becomes part of the story. A passing experience like "eyes open, no thoughts" or the peaceful relaxation of meditation is like a peashooter against a tank. The momentum of our collective drama has proved unstoppable.

If whole mind didn't exist, nobody could have invented it. We're all too stuck to see beyond the story we're tangled up in. But whole mind does exist; therefore, it must have a way of revealing itself. It is said of Lord Shiva in the Vedas that he has as many ways to show himself as to disguise himself. This is worth knowing. For every impulse of love, there is an impulse of fear. For every moment of clarity, there is a moment of confusion. The trick is not to favor the love and clarity over the fear and confusion, because the pendulum will inevitably swing back again.

The trick is to call upon the source of love and clarity, gradually come closer to it, and finally make it your own. (The New Testament had a better chance of being practical if Jesus had taught "Try to love your neighbor as yourself" and "See if turning the other cheek helps break the cycle of retaliation.") The only hope for escaping the traps of the conditioned mind is to use its better nature as a thread, following it day by day until your story fades away. Then, living in wholeness becomes natural.

# A Desire to Remove Desire

The endless cycle of good and evil, pleasure and pain, happiness and misery is like a bonfire that fuels itself. Yet just as the mind is riotously active on the surface and quiet in its depths, so is life. The drama is everything, but only at its own level. When you have a modest experience like "eyes open, no thoughts," the interference of mental activity ceases temporarily. What replaces it is more than quietude. By itself, a silent mind is no better, as far as negotiating the demands of daily life, than a car would be if the engine dies. There has to be something attractive within the experience of silence.

This something exists in the form of a faint presence, which we've already encountered as the sense of self. We are most natural, relaxed, free, and present when the sense of self is upon us. Sustaining this feeling of presence eludes us, however. I think it's fair to say that the glimpses of the true self that generate experiences of love, joy, peace, safety, and self-worth are the only thing that has brought light into the condition of being human. There is a natural desire to expand upon such experiences. In return, the true self has the power of attraction (known in Sanskrit as *Swarupa*, the pull of the Self).

Being silent and largely unnoticed, the sense of self cannot get us unstuck by itself. If you have practiced mantra meditation, mindfulness practices, or one-pointed awareness, which are all valuable techniques for bringing the sense of self to the fore, you know how stubbornly your mind returns to business as usual. You can train yourself with effort to display any spiritual attribute—loving-kindness, forgiveness, steady attention to God—but the sense of self cannot be trained. It simply *is*. Effort is not only useless but damaging. This concept was voiced by the venerable spiritual teacher J. Krishnamurti, when he said, "Can you discipline your mind to be free?" There's the rub. No matter how well we train our minds, including all kinds of spiritual training, freedom is a state where discipline is totally foreign.

I had that in the back of my mind when I said that the direct path must be easy, efficient, and spontaneous. If you want to be here now, which is the state of metahuman, you first have to stop trying. Meditation gives the experience of not trying, which surprises almost everyone when they first begin to meditate. The sense of self is brought into our awareness, and its presence can, in fact, be quite strong. It can turn into a state of ecstasy, a love affair with the Divine, as in this passionate outcry from Rumi:

> *Oh God, I have discovered love!*
> *How marvelous, how good, how beautiful it is! . . .*
> *I offer my salutation*
> *To the spirit of passion that aroused and excited this whole universe*
> *And all it contains.*

The usual experience of presence is pleasant and far short of passionate. We get a clue from Rumi, however, when he points to the power of desire. In his case, desire is an intoxication with divine love. Everyone has desires, and it is due to the force of desire that we wake up every morning with a will to see what the day is going to bring. If the day brings moments of awakening, then desire has served us in getting unstuck.

Since the sense of self simply *is*, you can't desire it directly. That would be like saying, "I wish I existed." What happens instead is that you trick desire into going where you want to go. (You spiritually hack into materialism, as a friend put it.) In Buddhism this comes under the heading of "Use a thorn to remove a thorn, then throw both thorns away." That teaching is famous and has aroused much commentary. It sounds simple enough to remove a splinter from your finger by pricking it out with a needle. The mental equivalent runs into difficulties. Since we are going to use desire to go beyond desire, these difficulties need a little explaining.

Originally, the Buddha seems to refer to thorns as deep pain that cannot be healed. One of the deepest sources of pain is fear of violence, which Buddha speaks of with personal candor. "Fear is born from arming oneself. Just see how many people fight! I'll tell you about the dreadful fear that caused me to shake all over. Seeing creatures thrashing like fish in shallow water, so hostile to one another! Seeing this, I became afraid" (Sutta Nipata).

We lean forward to hear what Buddha's escape from fear was, which conjures up the image of the thorn. He goes on, "Seeing people locked in conflict, I became completely distraught. But then I discerned a thorn, difficult to see, lodged deep in the heart. It is only when pierced by this thorn that one runs wildly in all directions. So if the thorn is taken out, one no longer runs but settles down."

In pure form, experiencing the thorn of pain in the heart spurs a person to find inner peace. The teaching is beautiful, but, like the New Testament teachings, it contradicts human nature. When people are locked in conflict, they are motivated to ramp up the conflict, ignoring the pain in their hearts, or, more likely, taking it to be a message that they must fight on. (To the credit of Buddhist cultures in Asia, the teaching of peace has brought fewer bloody conflicts over time, a record other faiths might envy. The Buddhist violence against Muslims in Burma proves that no religion is immune to violence, however.)

"A thorn to remove a thorn" can be used more psychologically as a tactic for healing. The goal is to substitute negative, self-defeating thoughts with positive, self-enhancing ones. Under the name of cognitive therapy, this approach relies on rationality to overcome emotion. For example, a patient might complain that nothing ever works out, that no matter how hard she tries, everything ends in defeat and disappointment. There's obviously a strong emotional charge behind defeatist thinking. When this becomes a habit, the first thought that comes to mind in a challenging situation is automatically "Nothing is going to work out. I know it."

The usual result is that, indeed, nothing good happens. There could hardly be a more perfect example of a self-fulfilling prophecy. A cognitive therapist would use a new thought to remove the old one (the thorn that removes the thorn) by pointing out that good things have happened to the patient. Life hasn't been a string of endless defeats. Therefore, the rational course is to think, "This situation may turn out well or badly. No one knows, so I have a chance to do my best to get a good outcome." It's not easy to replace pessimism with optimism, since emotions lie deeper than reason, and telling yourself that something good *might* happen gives pessimism all the opening it needs. With practice and guidance, habitual negativity can be softened, but human nature hasn't been made peaceful and fulfilled by the use of reason.

Finally, there is the second part of the teaching to contend with. "Use a thorn to remove a thorn, *then throw both thorns away.*" In the process of getting over an illness, you take medicine as long as it's needed, then you throw the pills away. But the same thing is more mysterious when it comes to the mind. If you use positive thoughts, for example, to replace negative thoughts, you still wind up as a positive thinker, just as embedded in your story as someone who is gloomy. It's better to live in sunshine than in shade, no doubt. Escaping the illusion is another matter.

Throwing both thorns away means that the whole scenario of pain and pleasure, good and evil, light and darkness, is discarded. This is taken to be Buddha's essential message. The diagnosis is accurate, and the picture of being healed—living beyond the grip of the drama—is inspiring. The challenge presented to the direct path is how to make inspiration a living reality.

## Clearing the Way for the Self

The answer, as I conceive it, is to use desire as the thorn, because we all have experienced its double nature. Desire demands to be fulfilled. If the desire is strong enough, as in passionate sexual attraction, it is enough to drive people mad. The thorn causes pain, because it is never enough to have one moment of fulfillment—the next desire arrives with a new demand to be fulfilled. What's the other thorn, the one that will end the pain? Fulfillment itself. Moralists who preach against desire, who call it the work of the Devil or, in Freudian terms, the work of the id, see with only one eye. Desire is as beautiful as it is treacherous. Sexual attraction leads to lasting, loving relationships and not just sexual assault.

I don't think anyone would deny this truism, and we can draw on it by using the thorn of desire to remove the thorn of desire. But desire for what? If spiritual traditions have held out the desire for love, peace, divine grace, forgiveness, and a ticket to Heaven or Nirvana, only to fall short in the end, what is left? The only answer I can think of is the desire to be real. The sense of self ultimately has that to offer, and that alone. There is no stick or carrot, only the promise that you can be here now, which is how you can experience reality free of all illusion.

In practical terms, the mind can't learn what it is to be real all at once. The direct path is still a path—it takes time. Life offers every imaginable reason to quit the path. Desire lures us to turn our interest elsewhere— the old story is very insistent. But you can cultivate an interest in getting real. It's not necessarily harder than cultivating an interest in wine or stamp collecting or Russian novels. All that's needed is curiosity and the flicker of desire (referred to in the Indian tradition as the spark that burns down the forest; one urge to escape the illusion will destroy the whole illusion eventually).

The direct path is unique because we are not taught by love, peace, compassion, or any other desirable quality our lives can bring. We are taught by existence itself. Only knowing what lies at the heart of existence will work. Contrary to Hamlet, "To be, or not to be" isn't the question. "To be" is the question, all by itself. If you ask it every day, your true self will hear and then present fascinating, ever-changing answers. Engaged in finding out what it means to be here, you transform your mind from its old conditioning into a vehicle that can reach the beyond. The conditioned mind has firmly decided what it means to exist, which is how we became stuck in the first place.

To get ourselves unstuck, once and for all, the direct path runs into a practical problem, which has been described in ancient Indian texts as "separating water from milk." In other words, everyday reality mixes the real and the unreal so they are inseparable. This is obvious when you look at the brain, which presents virtual reality only because the "real" reality makes us conscious in the first place. The worst movie ever made is projected on the screen with the same projector as a classic film masterpiece. Many versions of the direct path teach that we are like children watching shadow puppets on the wall. We believe in these images so completely that the truth escapes us—shadows can't exist without the light that makes them visible. Virtual reality only exists thanks to metareality.

To separate light from shadow sounds easy enough, and the direct path proves it. All you need to do is bring the sense of self to your attention. As a practical matter, nothing is easier or more natural. At any moment in the day, simply shut your eyes, take a few deep breaths to bring on a state of relaxation, and center your attention on your heart (or on your breath). It sounds strange, given the elaborate complexity of religion, that the entire spiritual journey is only this: relax and be aware of yourself.

By doing this, you subtly allow the "pull of the self" to become a real experience. If your intention is more serious, learning to meditate or making a commitment to yoga intensifies the experience of sensing the self—the mantras and yoga poses (*asanas*) were devised to give a different flavor or coloration of the self. We are already outfitted for this process. As children grow up, for example, they learn to identify the gradations of emotion. Hostility, outrage, irritation, and hate. Once you connect those labels with an inner feeling, it comes naturally to say, "I'm not angry, just frustrated" or "Don't go near him—he's a hater."

Gradations in consciousness can also be felt, but they don't have labels.

When you experience the sense of self as directed above, you go beyond the thoughts and feelings that can be labeled. None of them, even the most loving thought and spiritual feeling, is the basis of reality. Only the sense of self is irreducible. You cannot go beyond it or turn it into something more basic. Mental activity vanishes into a kind of nothingness—silent mind. According to the direct path, this is the source of everything "in here" and "out there." But here's the tricky part. Is "everything" practical? If someone showed you a blank piece of paper and said, "Here is every book" or a blank canvas and said, "Here is every painting," such a statement would not help you write a book or paint a painting. What you want is to write one book, not every book, paint one painting and not every painting.

Likewise, it seems unhelpful to say that metareality is "always, everything, and everywhere." We are attuned to specifics. For better or worse, we live our lives at *this* moment, in *this* location. It may seem hopelessly vague to tell someone as I've been saying to "Be here now." Where else are we, on Mars? But a life based on specifics is misguided, because the spell/dream/illusion contains thousands of specific things, so many that you cannot deal with them all or escape their grip. As a child in India, I heard simple versions of what it means to wake up to reality. In one tale, a fish is frantic with thirst, rushing around to find a drink of water. Even at four years old a child laughs at the silly fish who doesn't know it is surrounded by water. Only when you grow up does the seriousness of the paradox set in.

Once you have the experience of sensing yourself, which is available to each of us in our calmer moments, you can use inner silence to transform your world, one day at a time. This is done by looking at your mistaken notions about reality and rejecting them. Like a physicist sorting through all the objects in the subatomic world that aren't the irreducible building blocks of the universe, you examine your bedrock beliefs and sort them out on the way to arriving at your true source.

This task is made easy by two invaluable qualities of mind—insight and intuition. They are reliable guides when rational thought falls short. They are reliable even when emotions fall short. When people speak of "aha" moments, insight and intuition have cut through a tangled problem and gone directly to the truth. At such moments, when the light suddenly dawns, we exclaim, "I knew it, I just knew it was this way." The beauty of an insight is that it is self-justifying. You just know.

The sense of self is quiet and so subtle that few people bother to notice it, but within the self are things we just know. The direct path asks us to ferret out these bits of knowledge, so that they can be appreciated. The befuddled hero in a romantic comedy typically quarrels with the lady in his life until he knows—as she knew all along—that he loves her. We've seen this plot a hundred times, but it still delights us because insight, the moment of "Now I get it," brings clarity, and delight often follows.

Having insights about a relationship or a tough mathematical problem or the next chapter in a book you're writing is quite specific. Asking for insight into how life works is as nonspecific as it gets. The direct path comes to our aid by simplifying the mistaken beliefs we must see past in order to wake up. In fact, there are only a small handful to consider:

MISTAKEN BELIEF: I am separate and alone.
REALITY: The separate self is just a mental construct.

MISTAKEN BELIEF: This is my body.
REALITY: The body is part of the play of consciousness, which is
    universal, not personal.

MISTAKEN BELIEF: The physical world is the basis of reality.
REALITY: The physical world is one appearance that the play of
    consciousness takes.

MISTAKEN BELIEF: My life is bounded by time and space.
REALITY: Everyone lives in the eternal now, which is boundless.

At this point in the book, none of these mistakes comes as a surprise. Now we only need to get over them, which leads to a sorting-out procedure, as follows: At any moment when you have an experience worth noticing, pause and say to yourself, "This is the play of consciousness." So that the statement sinks in, it is best to have a sense of self along with it.

But that part isn't necessary. Nobody can separate milk from water (even a centrifuge only separates out the solids from the liquid), so they must be allowed to separate themselves.

This is a simple, effortless process, but one that requires patience. By reminding yourself that any experience is the play of consciousness, you bring reality to mind. You don't argue with yourself (or anyone else); you don't try to be good or wise or better than other people. You neither surrender nor strive. Instead, you sit for a moment at the intersection of awareness and reality, in a moment of reflection. The experience at hand is life at *this* moment, in *this* place. By consciously reminding yourself of what is actually true, you allow the "pull of the self" to bring you home.

Because we couldn't stand to be wrenched out of the conditioned mind all at once, the play of consciousness doesn't insist on anything. Life moves on like a river between its banks. When you say, "This is the play of consciousness," you aren't stopping the river or trying to steer it. You are sitting on the bank appreciating how the river flows. I hesitate to mention that this stance is known as *witnessing*, because once they hear the word, many people try to witness or get frustrated because they aren't witnessing when they want to. I sympathize. It sounds tempting to witness with calm detachment the death of a loved one, rather than being plunged into grief.

The river follows its own course no matter how we might wish it to move differently—it one day moves into grief, sorrow, pain, and suffering—the whole lot. Then it keeps moving. That's not simply sorrow, it's the inexorable flow that brings us to freedom. We only have to allow it, rather than struggling against it. Even struggle is just another aspect of the river, a small eddy that has no significance in the river reaching the sea. *Allowing* is one of the key words in descriptions of the direct path, but it is another word I hesitate to mention. People will wind up trying to allow, getting frustrated that they aren't allowing when they want to, and so on.

In pure form, the direct path is just experiencing the sense of self and letting it do its work without interference. Yet there is another stage, where silence and flow and the experience of "be here now" is over and done with. They aren't the goal, only a platform for launching the next stage. With the next stage, you no longer witness the flow—you *are* the flow. The play of consciousness and the self merge into one. Krishnamurti called this merging "the first and last freedom," a beautiful phrase. To make it more than beautiful, you have to experience it—let's start.

We will overturn the cosmos and everything in it.

# 12

# CHOICELESS AWARENESS

*Homo sapiens* is the only species that must try to be happy. Among all the mysteries that surround being human, this is one of the most intractable. For centuries our struggle against the enemies of happiness—violence, anxiety, despair, hopelessness, and depression—has been the price we pay for being conscious. We suppose, but do not know with any certainty, that other creatures don't experience these states. When it can no longer run with the pack, an old wolf lies down to die, defeated by age and the raw elements of Nature. A human being at any age can be warm, secure, and physically healthy, yet on the inside has given up on life.

Trying to be happy looks harder the more deeply you investigate human nature. Sigmund Freud, who devoted a lifetime to investigating what makes people tick, ended his days in London in 1939 as a refugee after Hitler brought Austria under German rule in 1938. As early as 1933, when Hitler first came to power, Freud's writings were banned, and as a Jew he was personally at risk.

To a disciple in England he wrote bitterly, "What progress we are making. In the Middle Ages they would have burned me. Now, they are content with burning my books." Insisting on staying in Vienna, Freud was finally persuaded to flee in the spring of 1938 when the Gestapo arrested and interrogated his daughter Anna. After paying an extortionate "flight" tax by which the Nazis ruthlessly stripped Jews of their money and property, the founder of psychoanalysis left for England in a state of pessimistic gloom, wondering as fear and terror blossomed like a poisonous flower if it was asking too much simply to prevent humans from committing murder.

Psychoanalysis has largely faded from the scene, replaced by drug therapies for mental disorders. No pill for happiness is in the medicine cabinet, however. In times of relative peace, people content themselves with a kind of happiness that skims the surface, finding love and pleasure when they come along and doing their best not to get trapped in the bad stuff.

Attempting to reshape human nature by probing what makes us tick hasn't succeeded. The direct path is about something different, sorting out what isn't essential in our lives. Pain and suffering are not essential. They are part of the drama we've constructed—a very stubborn part—but the drama can be left behind. There is a level of consciousness that transcends pain and suffering. The great insight of the direct path is that everything depends on your level of consciousness because everything *is* a level of consciousness. Metareality lies right next to everyday reality, being only a level of consciousness away. (Christian theology teaches that Heaven is where the soul finds eternal bliss, but long ago theologians abandoned the notion that Heaven is a physical place, despite the images carried around in our heads of angels sitting on clouds playing harps or of lambs gamboling through green meadows—that's one of the most common descriptions given when people come back from a near-death experience, claiming to have seen Heaven. The *Catholic Encyclopedia* defines Heaven as a state of being attained through grace. *State of being* and *state of consciousness* seem synonymous to me.)

So far we've taken the direct path through two realms of awareness, the body and the active mind. You might suppose that the journey has nowhere else to go. What more is there to life beyond the physical and the mental? There actually is something deeper, a department of existence that makes things turn out the way they do—good or bad, successful or a flop, a dream fulfilled or a dream denied. Here we discover that everything happens for a reason, but the reason is very different from the stories we tell ourselves. Pure consciousness organizes every event, including every possible event. It sees and knows all. From this viewpoint, everything happens for a reason because there is only one thing.

Yet the parts of life we work so hard to get right—job, family, relationships, morality, religion, law, politics—are not expressions of one thing. They are compartmentalized constructs. What happens in one can be alien or even anathema in another. It is against the law to take a life, but political leaders can win mass popularity by starting a war and taking an enormous number of lives. Nature ignores compartments at every level. A cell thrives as one thing, not a collection of parts. It eats, breathes, manufactures proteins and enzymes, divides, and expresses the knowledge compressed into DNA's double helix. To a cell, these are not compartments described in a medical textbook—parts don't matter. What matters is the life of the cell, which is one thing.

Can we get human life back to one thing, and, if we can, will it solve the problem of pain and suffering? The direct path affirms that this is possible. The one thing, as it applies to human beings, is pure consciousness, but giving it a label takes us further away from one thing, not closer. The great mystery of pure consciousness is that it has nothing that can be formulated, turned into a good way to exist, and trusted to bring happiness.

For example, in the Indian tradition pure consciousness is eternal, conscious, and blissful (*Sat Chit Ananda* in Sanskrit). This formulation led someone to think, "Aha, bliss is the key. What is happiness but bliss?" On that basis, "Follow your bliss" became a widely popular phrase. When put into practice, it had the advantage of pointing people inward, telling them that how they feel about their lives is more important than external rewards. Yet "Follow your bliss" is useless if you can't find your bliss in the first place, or if a tragedy or natural disaster has struck. It is worse than useless if, like Freud, you find yourself the target of unstoppable, malignant persecution.

The direct path doesn't rely on any formulations or fixed ideas about how to live. Confronted by the mystery of being human, it says, "Let the mystery solve itself. You are here to watch this happen. In that role, you are the living expression of the mystery." This is the closest anyone has ever come to living the one thing the way a cell lives one thing, which is itself. When distilled to its essence, the direct path asks you and me to dedicate our lives to the knowledge that we, in ourselves, are the mystery of life as it unfolds.

Beautiful words, but what do we do when we get out of bed tomorrow morning? The actions and thoughts that occupy our days aren't at issue. Tomorrow morning you will get out of bed and do what you want to do or have to do. Following you around with a video camera might detect nothing special about your everyday existence; it could seem like a typical existence followed by someone who has never heard of the direct path.

Yet on the inside the thing you have dedicated your life to will be very different. In every culture where people have seriously investigated one thing (a philosophy technically known as "monism"), the same dedication was present. Whether it arose in ancient India, Persia, Greece, or China, monism directed the mind away from the countless details of life toward one thing. That's what this chapter is about. Despite the allure of diversity, the infinite variety of things held out by the physical universe, the mystery offers no solution there.

When you reduce life to what is truly essential, diversity drops away and unity increases. Your consciousness starts to reshape itself around one thing. Waking up is one thing; enlightenment is one thing; allowing the mystery of existence to unfold through you is one thing. These descriptions are limited, however. What we're really after is the experience of one thing. Once this is attained, everything else follows, not just the end of suffering, but access to the infinite potential that wants to express itself through us.

This is the state I'll call "choiceless awareness." It is the very opposite of struggle. Instead of always doing, you practice the art of not-doing. Instead of trying to decide if X will make you happier than Y, you let the choice happen on its own. When we hear about the advantage of letting go, living in the now, and allowing life to flow, these possibilities become real in the state of choiceless awareness.

Choiceless awareness is the final stage of waking up. It brings you to a place where the next thing you want to do is the best thing for you. In such a state, pain and suffering end because they bear no relationship to the life you are consciously living. The mystical Indian poet Kabir saw the situation clearly:

> *I asked my heart,*
> *Where are you bound?*
> *There's no traveler ahead of you*
> *Or even a road.*
> *How can you get there,*
> *And where will you stay?*

The road, in Kabir's imagery, is about the stages of leaving, going along, and arriving, which is how we all live. We start an action, either a small one like getting orange juice out of the fridge, or a large one like getting married or finding a job. The action begins, goes along, and ends. Kabir sees that the heart—his word for the soul—cannot find fulfillment this way. He sees another way instead:

> *Be strong, O my heart,*
> *Put your fancy away,*

*And stand where you are*
*In yourself.*

Let's see if we can manage to do just that.

# The Art of Not-Doing

Choiceless awareness sounds alien in the modern world, where life is viewed as nothing but choices, and happiness is thought to be reached by making good ones instead of bad ones. It's hard to realize that the best choices make themselves. The rationale behind not-doing is simple: one thing has nothing to do but be itself. Therefore, we have nothing to do but be ourselves. This is the essence of not-doing.

It's not a new idea. In the Taoist teaching of ancient China not-doing is *Wu wei*. The concept arose as far back as five to seven centuries BCE. In Buddhism the teaching is found in a sutra like the following: "Happiness can be accessed directly in meditation by abstaining from conscious mental activity." In Christianity, the leap of faith that originated with the Danish philosopher Søren Kierkegaard was his attempt to carry out the central concept in the Sermon on the Mount, that all things can be left to Providence. In a leap of faith, a believer lets God take over, the ultimate experiment in trust.

The absence of struggle is common to all teachings of not-doing, but the passage of centuries hasn't made it less mysterious. Not-doing makes no sense to the logical mind. Of course we have lots of stuff to do—we barely find enough hours in the day to accomplish them. But what did you do today to produce new red blood corpuscles, the lining of your stomach, and the outer layer of your skin? These are parts of your anatomy that must be constantly replenished because blood, stomach, and skin cells have a life span of only a few weeks or months.

For that matter, what will you do today to exchange oxygen for carbon dioxide in your lungs, without which life is extinguished in a matter of minutes? Already the vast proportion of your physical life takes care of itself. Existence is divided between things we do and things that take care of themselves. Therefore, some part of you is already living without making choices. To begin to see the value of choiceless awareness, stop assigning so much importance to your everyday choices, which are mostly governed by habit. In the long run, habits keep you from finding renewal. Even worse, while entangled in virtual reality, you may fall prey to the feeling that life is incredibly unfair. You can be smart, gifted in all kinds of ways, ambitious, and celebrated for your achievements, yet there's no guarantee that you won't suffer the fate of Mozart.

In the late summer of 1791 Mozart went to Prague, where one of his operas was the centerpiece of festivities for the Austrian emperor. He began to feel ill but returned to Vienna and tended to the premiere of *The Magic Flute*, one of his greatest masterpieces, on September 30.

By then his health had taken an alarming turn—he experienced swelling, vomiting, and severe pain. His wife and doctor did their best to nurse him back to health as Mozart's mind turned to writing a new commission for a requiem, which he never finished. Occupying himself with his musical genius was in vain, and an hour after midnight on December 5, Wolfgang Amadeus Mozart, the most precious of musical minds, died of still-unknown causes at the age of thirty-five. There were heartrending scenes toward the end of his life of him sitting up in bed to sing ditties from *The Magic Flute* with friends. Few who love classical music consider Mozart's death anything but a great loss.

I don't bring up this sad story, which reached many nondevotees of classical music through the 1984 movie *Amadeus*, to mourn a death or to speculate about what modern medicine could have done to save Mozart's life. (Nearly twenty causes of death have been conjectured, including rheumatic fever, kidney failure, and blood poisoning. Most of them are infections curable today with antibiotics.) I merely want to point out that the greatest among us is still controlled by unseen causes. The drama we cling to simply does what it does, on its own terms. The spell/ dream/illusion mysteriously generates unpredictable events that cannot be fathomed.

How do we get beyond the sense that we are helpless before random chance or invisible Fate? Without a doubt there are moments when hidden links are revealed. The notion of synchronicity has become popular for describing a special kind of coincidence. You think of a person's name, and a few minutes later that person calls. You want to read a certain book, and, without being told, a friend brings you the book. *Synchronicity* is defined as a meaningful coincidence, setting it apart from random coincidences that have no meaning, like seeing the same make of car as yours show up beside you at a stoplight.

Metareality has a way of sending us messages now and then that contradict our fixed views. Synchronicity is one kind of message. It says, "Your world is disorganized, but reality isn't." In a fleeting glimpse, we see that a higher intelligence can organize events so that meaning pops up unexpectedly. Moreover, a synchronous experience defies our limited conception of cause and effect. Instead of A causing B, the two are invisibly meshed. If you think of a word and the next minute someone on TV says that word, you didn't create the coincidence. But *something* did.

You can feel your physical body and listen to your thoughts, but the something that organizes all events cannot be sensed. Yet it is here and now. The following exercise will show you what I mean.

# Exercise: I'm Still Here

Wherever you are now, look around the room and take an inventory of what it contains. Notice the furniture, knickknacks, books, even the windows and the view outside.

Now close your eyes and see the room mentally. Begin to remove its contents. (You can open your eyes to refresh your memory if you need to.) See the furniture go poof! And no longer exist. Strip the room of the knickknacks and books, then the windows and the view outside. You will be left standing in an empty box. If you have a mental image of your body standing in the room, send it away, too.

In one final stroke, remove the ceiling, walls, and floor. You will be standing in a space with no objects. It's hard to have an image of nothing, so you probably see white light or blackness. Notice, after removing everything, that you are still here.

Now reverse the process. Bring back the floor, walls, and ceiling. Reinstate the furniture, knickknacks, and books. Replace the window and the view outside. Notice that you are still here. Nothing about the room, whether full or empty, changed you.

To prove that you are still here, try to get rid of yourself. See yourself standing in the room when it is full or the space left behind when the room has disappeared. In either state, can you remove yourself? No. "I'm still here" is untouchable, because you are the one thing.

This exercise helps remind you of your true status. If you find yourself in a busy place, full of distractions or stress, take an inventory of everything you see. Close your eyes and empty the place until there are no objects, people, or room. As you arrive at nothing but emptiness, you are still here. This gives you a feeling of blissful self-sufficiency. To know that you are imperishable, *right this moment*, is the ultimate validation that you are here now.

You have discovered your being. After that, everything you say, do, or think is a passing shadow. Simply by being, you have mastered the art of not-doing. The next stage is to let one thing show you what it is capable of.

# Practical Immortality

When you see yourself as the unmoving point around which everything moves, you are in a special state of awareness. We can call this the state of *practical immortality*. Countless people already live such a life if they are part of a religion that believes in a personal God. Believing that God is watching and judging them, devout believers are basing their lives on an immortal being. The ultimate reward from such a being is going to Heaven, where the believer gets to join in immortality.

In societies dominated by traditional religion, rules of morality tend to be rigid and narrow. Dogmatic religions nail a set of rules to the wall saying, "If you want to know what God wants, obey these rules." The rules can be as basic as the Ten Commandments and the Golden Rule or as complex as the hundreds of daily instructions that pertain to the life of an orthodox Brahmin in Hinduism. But living by the rules depends on believing that there was a line of communication between God and whoever delivers the rules here on Earth. Religions tend to agree about what constitutes a devout life (i.e., trying to please God and obey the rules), but they disagree on which messenger to trust.

The direct path doesn't propose that kind of practical immortality. There are zero rules or even injunctions. Instead, there is a path that arrives at waking up. Once you are on the path, you must communicate with immortality yourself, as best you can. Moses, Jesus, and Buddha aren't going to help you (although, because there are no rules, if you want to follow a religion, the choice is yours). To begin to communicate with immortality, the first step is to believe that you can, which isn't easy.

For our entire lives, all of us have been taught how to live in the state of practical mortality. Everything is gauged according to how it fits in between birth and death. People who squeeze the most pleasure out of life, master many skills, work hard and play hard, travel to many places, amass money and possessions—they are the winners in the scheme of practical mortality. At its most basic, however, everyone lives in the shadow of mortality. You see this in our worship of youth, our obsession with fending off aging, our fear that our bodies might get injured. Protecting the body lies at the heart of getting rich—money offers a surrogate form of invulnerability. The rich and powerful come close to being invulnerable to life's hardships. (Yet nothing is more untrue than the saying, "It's better to die rich than poor." Finality is finality, no matter how much money you amass and throw around.)

In this context, how can you believe that you can communicate with immortality, much less turn it into a way of life? All you actually need is the insight that you are doing it already. Wholeness—the one thing we've been talking about—is immortal. It communicates with mortal life by emerging into the physical world as time, space, matter, and energy. Physics has no problem with that assertion. The precreated state or quantum vacuum is not bound by time or space. It contains no matter and energy but only the potential for matter and energy.

The direct path translates this basic fact into something profoundly human. When you do the exercise in this chapter of emptying a room until you are left with a blank slate, you set the stage for filling the room up again with anything you want. As a symbol of human creativity, the exercise is simple, but when you look around you, the entire world has been constructed the same way. Colors, textures, tastes, smells, and the like are all symbols for a quality human beings desired to have in virtual reality.

We keep at this project, taking a possibility, plucking it out of its state of potential, and then making it real. We embrace dogs for their human traits—loyalty, friendliness, obedience, and so on—while in return, through a sort of bleed-through or osmosis, the primal Siberian wolf from which all domesticated dogs are descended has been given an evolutionary makeover. Humans consciously selected the traits we consider desirable and excluded those we don't wish to see. Adorable eyes are a very desirable trait, and a golden Labrador puppy as young as eight weeks instinctively gazes into its owner's eyes, forming a bond that we instantly recognize. (Wolves, on the other hand, retain a feral glare with no connection to humans.) The dog is a human creation in the flesh as well as in our creative imagination.

Hybridizing dogs and cats, like manipulating the genes of corn and wheat to resist pests, is physical manipulation, but the force behind it, imagination, isn't. Using our imagination is one way you and I and everyone else are living the mystery. No one created imagination, just as no one created creativity. No one created evolution, either. When you get closer to the source, where consciousness in all its diversity begins to narrow down to one thing, it is evident that being human depends on contacting the immortal domain constantly, not just every day but every second.

Diversity is a blinding spectacle. Seven billion people with the capacity to see millions of colors, write countless melodies, speak infinite combinations of words, and pursue endless dreams and obsessions—this whole panorama is diversity. Being human cannot be confined to any set of rules, whether handed down by divine authority or created by human lawgivers and enforcers.

What makes us human is invisible and impossible to codify. To be human, you must know how to pay attention. You must understand what it means to remember something, hold it in your mind, and return to it whenever you need to. You must be able to carry through with an intention. These things are so basic that we barely notice them, but too often the basic instructions have gone awry.

The average mind is *restless*, unable to stay still for more than a brief moment.

The average mind is *shallow*, unable to reach beneath the surface of unending mental activity.

The average mind is *without purpose*, unable to bring its intentions to fruition in a meaningful way.

These three problems are interlinked, and they bring about the struggle that most people know all too well. If it were natural for the human mind to be restless, shallow, and without purpose, it would be useless to speak of metahuman as anything but an empty ideal. But, in reality, the mind by nature is restful. It is capable of going deeply into its own awareness, and it can find a higher purpose. Practical immortality brings this truth to light, telling us what is normal and natural in our mental life.

When you are in contact with one thing, other basic things become clear. You cannot be human without the ability to make symbols and recognize what they mean. A red stop sign is a symbol that tells cars to halt at an intersection. *Red* has zero connection to *stop* until human beings assign it that meaning. The word *tree* symbolizes a class of tall plants with wooden trunks, but the connection is totally arbitrary. *Arbre* will do in French; *Baum* in German. Yet having created a symbol, we can become imprisoned by it, the way that flags trap people into nationalism, money into greed, and religious rituals into dogmatism.

The freedom to create symbols without being ruled by them is part of practical immortality. The Sufis say that everything in the world is symbolic, which seems true to me. Virtual reality is a three-dimensional symbol of how *Homo sapiens* wants the world to be. The things we say in words are symbols of deeper things that are beyond words or even thoughts. If you try to unravel everything implied by the word *love*, for example, the thread of meaning will lead in every direction. "I love X" can be assigned to all desires, all wants and wishes. Then the opposite of love, whether it is hate or fear, becomes symbolic of the undesirable.

As part of going beyond your story, it is necessary to stop falling prey to the lure of symbols. You begin to see that they are like counterfeit money. *Love* can be the word used to justify domestic violence, obsessive jealousy, stalking, or attacking someone else who desires the same person you desire. The same slipperiness afflicts a word like *peace*, which can be used by a country that sells weapons of mass destruction in the name of keeping peace among nations or appeasing tyrants to prevent them from murdering their own people.

The ultimate rationale for practical immortality is the same as the rationale for not-doing. You let wholeness unfold on its own. You stop forcing, struggling, and interfering. It's unfortunate that the teachings of not-doing, whether in Buddhism, Taoism, or Christianity, have acquired the reputation of being mystical. If people simply stopped forcing, which isn't mystical in the slightest, daily life would improve immensely. If we stopped doing a lot of things we know are bad for us, not-doing would become a way of life everyone would want to follow.

Why don't we? Because everything involved in practical immortality—paying attention, using your intention to get somewhere, fulfilling a purpose, dropping your story, ending the tyranny of symbols—depends on a state of consciousness you have to arrive at. Virtual reality is all about stories and symbols. Metareality isn't. That's why a path is needed to get from one state of consciousness to a different one. What choiceless awareness teaches us is that the path isn't about what you think, say, or do. You unfold your potential, which isn't a matter of thinking, saying, or doing. It happens on its own.

The truth is that no one has to worry about communicating with the immortal, because the immortal is constantly communicating with us. It won't let us go, ever. A message from metareality enters our awareness, and it sticks. The message can be anything once it is translated into daily life. Look around you. No one invented attention, intention, love, intelligence, creativity, and evolution. Yet these things are here, now and forever. They never let go, no matter how much we misuse them. One thing has us always in mind, which is why we have it in mind. Neither side has a choice.

# 13

# ONE LIFE

When we are whole, the world will become whole. That would be an amazing change, because, as things stand, we are divided and so is the world. This state of affairs goes beyond the never-ending conflicts that make the news. There's a deep fracture at the very core of being human. We call ourselves mammals, and yet most people believe they have a soul. We set ourselves apart from Nature, exploiting it without considering the consequences. As caretakers of the planet, we are also its worst threat.

A shift is already being sparked in collective consciousness, however. One of the most heartening signs of this is seemingly trivial, an online video about a grateful octopus, now seen by almost twelve million viewers. It begins on a beach in Portugal with a man, Pei Yan Heng, strolling along the sand. He spots a small octopus stranded out of the water. Pei takes out his smartphone to film the creature. Presumably tossed onshore by a large wave, the octopus looks shriveled and near death. In a kind gesture Pei gathers the octopus into a plastic cup, carries it back to the sea, and releases it.

The octopus immediately begins to revive. Its eight arms spread out (experts tell us that *tentacle* isn't the proper term), and it changes to a healthier color. Typically, an octopus is shy and scuttles away from any approaching threat, a necessary tactic for such a baggy, soft-bodied animal. But instead of fleeing, the saved octopus approaches Pei's boots and places two arms on them, resting for several seconds before moving away without haste. Soon the "grateful octopus" entered popular culture when the video went viral. You might suppose that this was an example of human sentimentality, but then, there is no proof that the rescued octopus *wasn't* grateful. Is there any way to tell?

The conventional answer is no. This can be a hard no or a soft no. The hard no holds that only human beings are conscious. The soft no holds that humans are the only fully conscious creatures. This leaves a little wiggle room for big-brained mammals, like porpoises, elephants, and the great apes. The soft no has held firm for a long time. But once you understand that consciousness is the source of creation, the road to yes—an octopus *can* feel gratitude—is opened.

From the perspective of metahuman, nothing is alien, however. There is only one reality, governed by one consciousness. There is only one life, too, despite our distinctions between smart chimpanzees, stupid lizards, and totally unconscious bacteria. An urgent need at this moment when the Earth is in peril is to evolve to metahuman for the sake of all living things.

# One Life, and One Life Only

In a January 2014 article in *Scientific American*, the highly regarded neuroscientist Christoph Koch made inroads against the "no" position by asking if consciousness is universal. He is very persuasive when he points out that animal intelligence isn't primitive. Not only that, it isn't correlated with brain size or even possessing a complex nervous system. "Bees can fly several kilometers and return to their hive, a remarkable navigational performance," Koch points out (not only remarkable, I would add, but something human beings lost in the woods are incapable of). "And a scent blown into the hive can trigger a return to the site where the bees previously encountered this odor."

Koch links this trait, called "associative memory," with the famous moment in French literature centered on a cookie known as a madeleine. The massive seven-volume novel *À la recherche du temps perdu* (*In Search of Lost Time*), by Marcel Proust, begins with a flood of memory occasioned when the narrator dips a madeleine in a cup of tea, a gesture from his childhood. This experience of associative memory Koch also ascribes to bees, a lowly form of insect life. But we can find a wealth of other examples. Staying just with bees, Koch points out the following:

> [They] are capable of recognizing specific faces from photo-
> graphs, can communicate the location and quality of food sources
> to their sisters via the waggle dance, and can navigate complex
> mazes with the help of cues they store in short-term memory (for
> instance, "after arriving at a fork, take the exit marked by the
> color at the entrance").

The upshot, Koch says, is that consciousness cannot be walled off in arbitrary ways just because a life form looks biologically too simple to be conscious. With arms wide open, he declares, "All species—bees, octopuses, ravens, crows, magpies, parrots, tuna, mice, whales, dogs, cats and monkeys—are capable of sophisticated, learned, nonstereotyped behaviors." This gets us a long way from "no," only human beings are conscious, to "yes," consciousness is universal.

The grateful octopus was acting out a human gesture. To see this isn't sentimentality or fantasy. Koch believes that if we weren't so prejudiced, we'd see that animals behave constantly in ways that would be called conscious if the same activity were displayed by a person. A dog's look of love toward his master, his distress if his owner is absent, and the grief he feels if his owner dies are conscious traits being expressed through another life form. Yet our prejudice is hard to overcome because it serves our selfishness. *Homo sapiens* has ancient hunting stock in it. We kill and eat a great many animals, and it salves our conscience to see them as lower life forms, deprived of mind, will, and freedom of choice.

Everything that makes other life forms alien in our eyes is arbitrary. No creature looks more alien than an octopus. Among the three hundred octopus species, which first appeared at least 295 million years ago according to the oldest fossils, the largest types resemble the smallest in having two eyes, eight arms, and a beak centered where the arms meet. Blown up to a large scale, as in the giant Pacific octopus, which can reach a weight of up to six hundred pounds with an arm span between fourteen and thirty feet, those eight arms and snapping beak seem monstrous. But like *Tyrannosaurus rex* or a great white shark, the giant Pacific octopus isn't a monster in its own eyes. In the play of consciousness, the octopus occupies the same cosmic status as *Homo sapiens*. It is alive and aware of itself and its surroundings.

There's abundant evidence to support this claim. In her 2015 book, *The Soul of an Octopus*, naturalist Sy Montgomery closes the gap between people and mollusks in startling ways. In a section that begins, "Octopuses realize that humans are individuals, too," she relates how distinctly an octopus can make friends and enemies. In the mildest example, one keeper at the Seattle Aquarium was assigned to feed the octopuses while another touched them with a bristly stick. Within a week, at first sight of the two people, most of the octopuses drifted toward the feeder.

But their ability to relate to specific humans grows much more mysterious. A volunteer at the New England Aquarium earned the dislike of a particular octopus named Truman for no apparent reason. Whenever she came close to the tank, Truman would use its siphon (a funnel on the side of an octopus's head that propels it through the water) to spray a blast of cold seawater at her. The volunteer went away to college but returned months later for a visit. Truman, who hadn't squirted anyone in her absence, immediately soaked her with a blast from his siphon at first glance.

Montgomery relates in depth the idiosyncratic behavior of octopuses in captivity named Athena, Octavia, Kali, and others, making them almost as individual as people. Her argument for their similarity to humans is ultimately physical. After all, she writes, we share the same neurons and neurotransmitters. But even though octopuses have unusually complex nervous systems for an invertebrate, their anatomy doesn't resemble the human nervous system. The majority of octopus neurons are located in their eight arms, not in their brains. Each arm can independently move, touch, and taste (the suckers that line each arm are locations for the sense of taste) without needing to refer to the brain.

Anatomy cannot explain how octopuses recognize people and remember their faces. Disliking a bright light at night that disturbed its sleep, one octopus aimed a jet of water at it and burned it out through short-circuiting. Dissecting an octopus's nervous system doesn't explain how such a tactic was devised (in the wild octopuses don't squirt water above the surface of the seas). It would seem to be an act of creative intelligence.

My contention is that existence *is* consciousness; therefore, no animal ability is astonishing (except in our one-eyed view), because every life form expresses traits that belong to pure consciousness. These traits wake up, as it were, emerging into the physical world according to each creature's evolutionary story. The grateful octopus wasn't being like a human. We could say with equal justice that when we are grateful, we are being octopus-like. Both views are one-eyed.

This book has been making the case for waking up, but being awake isn't the end—ahead lies cosmic consciousness. I am using the term the way others use *supreme enlightenment* (known in Sanskrit as *Paramatma*). If metahuman is the awakened state, think of it as crossing a threshold. There is a vast new territory to explore beyond.

# Cosmic Consciousness

Cosmic consciousness doesn't give a little bit of itself to an amoeba, more to bees, still more to octopuses, and finally the grand prize to *Homo sapiens*. In a hologram, a fragment of a laser image can be used to project the whole image—with only *Mona Lisa*'s smile, the entire painting can be projected. Hologram technology can even simulate a statue or a living person in 3-D from a laser image in two dimensions. Cosmic consciousness does this on a huge scale—the entire universe—using merely the *possibility* of a cosmos. Therefore, it's not quite true that something is created out of nothing. The physical universe sprang from a conception in cosmic consciousness that unfolded in material form. Pure consciousness is not nothing.

This capacity has been inherited by human beings. If I say, "Imagine the Eiffel Tower" or "See the Statue of Liberty in your mind's eye," it only takes the name of those monuments for you to see them in totality. A name doesn't have three dimensions; in fact, it has no dimensions, being just a verbal tag for a concept. The Statue of Liberty is the concept of freedom transformed into a work of art. But *liberty* can also produce completely different manifestations, such as revolutionary wars or an antiwar movement. Concepts are constantly shaping and reshaping events, civilizations, and the human world in general.

You are living in a world that consists of ideas blown into three dimensions. As usual, the great minds got there before us. More than two thousand years ago Plato argued that everything in the world originated in abstract universal ideas, which he called "forms." Leap ahead two millennia, and here is Werner Heisenberg: "I think that modern physics has definitely decided in favor of Plato. In fact the smallest units of matter are not physical objects in the ordinary sense; they are forms, ideas which can be expressed unambiguously only in mathematical language."

If the elementary building blocks of matter and energy are conceptual, then the universe itself is bubbling up from a set of ideas or forms, too. This particular set of ideas that became our home universe could have other variations, some of which would be inconceivable to the human mind. One feature of the multiverse, if it actually exists, is that billions of other universes may be operating on totally different laws of Nature from ours. A law of Nature is simply a mathematical model, and mathematical models are concepts.

Let me interject a personal note here. When I first encountered the quantum, which led to a book, *Quantum Healing*, I was thrilled that physics was in accord with profound insights from India. *Maya*, the Sanskrit word usually translated as "illusion," refers to virtual reality, and the doctrine of Maya holds that the illusion is merely a concept. The parallels went even deeper. Heisenberg held that Nature exhibits a phenomenon according to the questions we ask of it—in other words, the qualities of time, space, matter, and energy are extracted from the quantum field by the observer. In ancient India, Maya originates through the participation of humans seeking confirmation of our inner beliefs. In both cases Nature is showing us what we want to see.

I was thrilled by the prospect that the inward path of the ancients and the outward path of modern science had arrived at the same reality. So it came as a shock to discover that contemporary physics has largely turned its back on the inspired quantum pioneers. As a professor at Cal Tech told me, "My graduate students know more about physics than Einstein ever did." This advance in technical knowledge has been tremendous, but does it justify throwing out what the quantum pioneers understood about reality?

Einstein at least recognized the danger when he remarked, "So many people today—and even professional scientists—seem to me like someone who has seen thousands of trees but has never seen a forest." To correct this short-sightedness, Einstein advocated that scientists acquire a broad outlook from philosophy, which he considered the mark of "a real seeker after truth." In the twenty-first century, alas, fewer forests are being seen today than ever as every science becomes more specialized and fragmented. You can spend an entire career in physics focusing on a single concept like eternal inflation or a single elementary particle, like the Higgs boson.

Cosmic consciousness sounds like something very far away from how we use our minds in day-to-day life. In reality, however, each person's mind is projecting cosmic consciousness all the time. Your mind is a fragment of cosmic consciousness, yet, as in a hologram, a fragment is enough to project the whole. The following exercise will help bring this insight home.

# Exercise: Surfing the Universe

Close your eyes and imagine that you are standing on a beach, watching the surf roll in. When you have this image firmly in your mind's eye, start transforming the waves in various ways. See them become bigger, swelling into the monster waves that world-class surfers ride. See them shrink to small chop. Make the waves turn different colors—red or purple or neon orange. Place yourself on top of the waves, balancing without a surfboard as you ride to shore. If you wish, you can invent your own transformations. Perhaps a mermaid emerges from the waves, singing her siren song. You get the idea.

When you made these creative changes to the surf, reflect on what was happening. You didn't thumb through a catalog of possibilities. You were free instead to let your imagination roam. Two people doing this exercise would come up with different creative choices. The possibilities are unlimited and not bound by any rules. Nothing stops you from turning the Pacific Ocean into pink Jell-O. It makes no sense to claim that these creative possibilities are stored in the atoms and molecules of your brain cells. You made conscious choices without precedent, building a unique chain of creative thoughts.

But even if all seven billion people on the planet performed this exercise, they would be doing only one thing—transforming possibility into reality. This one thing is occurring all the time, and it is enough to create the universe. In the spring of 1940, one of the most far-seeing physicists in modern times, John Wheeler, telephoned another far-seeing physicist, Richard Feynman.

"Feynman," Wheeler exclaimed, "I know why all electrons have the same charge and the same mass."

"Why?"

"Because, they are all the same electron!"

This startling notion, which became known as the One-Electron Universe, sank into Feynman's imagination, although, as he recalls, he didn't take it seriously enough at first. When we look at the physical world, a huge number of electrons exist—trillions send electrical charges through your household current every second. Each electron traces a path in time and space, known as a "world line."

Wheeler proposed that a single electron could zigzag all over the place, creating a tangle of world lines. It's a fascinating alternative to many electrons creating many world lines. Now let's translate this into human terms. Instead of many electrons, substitute many observers, each with his own eyes. On planet Earth there would be over seven billion observers. However, those billions of observers express the ability to observe, which is one thing. So it's entirely plausible that we inhabit a "one-observer universe." It's like saying "All humans draw breath" without having to count how many humans are breathing. Such is the perspective of cosmic consciousness. I didn't choose the image of waves pounding the shore by accident. The ancient Indian seers pointed to the sea and said, "Each wave is an outcropping of the ocean without being different from the ocean. Do not be fooled by your individual ego. You are an outcropping of cosmic consciousness without being different from it."

*Homo sapiens* is the only creature that can choose which perspective to take. We can be separate waves of one ocean. The only difference between a one-electron universe and a many-electron universe is our perspective. Both are as real as we decide they are. Or, to put it more strongly, both are *only* as real as we decide they are. Standing at the pivot of making this choice, we stand at the pivot of creation. Only one thing is happening: possibility is becoming reality. John Wheeler was also responsible for saying that we live in a participatory universe. I'm only expanding on the same idea. A participatory universe offers infinite choices; the one thing you can't choose is *not* to participate.

Once you are in the game, how you play it is entirely open to you. Humans can look upon creation and explain it any way we choose. Why are there cold viruses, elephants, sequoias, and mice in the world? Some may say God created them on purpose, while others believe that they emerged from the quantum vacuum through random processes that took billions of years to come to fruition. The most radical explanation is that *Homo sapiens* added everything we desired to our virtual reality. Each explanation is simply a different story. Beyond stories, cosmic consciousness is creating from within itself. Stories are postcreation; cosmic consciousness is precreation.

# The Causeless Cause

Our role as creators of reality imposes a heavy burden if we view it from the limitations of human nature. For many centuries the whole thing could be left to God. The medieval mind, for example, made God the origin of everything in Heaven and on Earth—Thomas Aquinas, the greatest of medieval theologians, presented God as the "first mover" (*primum mobile* in Latin). God alone had the knowledge to create the universe.

Because perfection is a divine attribute, God must have set creation in perfect motion, while in the fallen world everything that is in motion, even a beating heart and surging ocean waves, is an imperfect representation of God's work. When Adam and Eve fell, so did Nature. The first humans were driven out of a perfect natural world into an imperfect one. The Garden of Eden gave way to a hostile wilderness.

In the *Divine Comedy*, which stands as the most complete reflection of medieval cosmology in literature, Dante reached for a visual image of divine perfection that his readers could grasp. As the website Dante-worlds describes it, "In the Primum Mobile ('first mover')—the swiftest, outermost sphere that imparts motion to the other spheres—Dante sees nine fiery rings whirling about a central point of intense light."

These nine fiery rings are angelic orders, because in Dante's religious worldview there had to be perfected beings assigned to keep creation going. Otherwise, God would have to be engineering everything, an impossibility when he was the Unmoving Mover by definition. (*He* is archaic and not correct in Hebrew, but I'm resorting to the masculine for convenience, since *he/she/it* is cumbersome.) When referring to God, the medieval Christian mind couldn't violate—or escape—divine perfection.

That obsession survives today, but in a different guise. With no perfect creation to dream about, we are left with our own imperfections. We feel as bewildered and confused as the biblical Adam and Eve. We feel guilty about despoiling the planet and yet cannot help ourselves, even as Nature crumbles before our eyes.

This book has proposed that creation unfolds from pure consciousness. There is no divine artist with a picture in mind. There is just creation evolving without end. The process has no one story line; it embraces all story lines. It has no morality. Tragedy is as fascinating to the human imagination as comedy, which is why we keep creating both. (Shakespeare presented the whole panorama to his audience standing in the pit of the Globe Theatre, and Hollywood keeps the show going.)

The evolution of consciousness is the only explanation for creation that holds everything together. It has the advantage of no boundaries. The miraculous stands on the same level playing field as the mundane. At this point I am going to go out on a limb. If you go to YouTube and enter three search words, *Lourdes levitating Host*, you can watch a filmed miracle that occurred in Lourdes, France. As one informed online commentator explains it:

> In 1999, during a Mass celebrated by Cardinal Billé, then archbishop of Lyon, the Host began to levitate just above the paten [the plate used for Eucharist] from the moment of the *epiclesis* until the elevation. The prodigy was filmed for broadcast and a clip of it is making its way around the internet. At the time, the French bishops decided to keep it quiet. Recently, it was brought to the attention of a cardinal in the Curia, who took it upon himself to verify the origin of the clip and ask the current archbishop of Lyon for the position of the French bishops on the matter. This cardinal in turn passed it on to the Holy Father. He is concerned that certain bishops were too quick to put the lid on what seems to be an authentic sign.

The existing video is blurry, but it shows what the commentator describes. During this Mass a large Host was used, about the size of a dinner plate. The levitation, which lasts several minutes, ends with the elevation of the Host, when the archbishop lifts it up to display to the congregation. The levitation, if that is what we are seeing, raises the Host only an inch or two into the air.

I don't know who is in a position to rule the footage real or a clever digital hoax, but for me the issue isn't about miracles. It is about what human beings are willing to allow into the acceptable picture of reality. To date, millions of people have seen the video of the levitating Host, and their responses cover the spectrum. Most people I know are momentarily impressed; others question the blurry images. A few get a strange expression on their faces, as if they were Horatio and Hamlet had just said, "There are more things in heaven and earth, Horatio, / Than are dreamt of in your philosophy."

You could say that Hamlet is accusing his friend of not dreaming deeply enough. Miracles are similar reminders. The levitating Host may be explained away one day—after all, antigravity exists in theoretical physics. It might be exposed as a fraud or simply sink into the morass of forgotten experience. Even so, something important happened. A bit of strangeness was allowed to enter our collective dream. It takes only a spark to burn down a forest. You never know which strange bit will dispel our collective dream.

The time for waking up never grows short. Waking up takes you beyond the bounds of time. Yet it's hard not to feel the pressure of disaster the closer it approaches. There is such a thing as a storm so powerful it occurs only once every five hundred years, but by the reckoning of meteorologists, twenty-six such storms have occurred in the past decade. If we are going to make cosmic consciousness matter, we cannot wear rose-colored glasses. Those storms, and the human misery they generated, were allowed into virtual reality. Many things have entered virtual reality to make life nightmarish.

The average person isn't prepared to accept responsibility for the spell/ dream/illusion we are entangled in. The accumulation of greenhouse gases can be explained as divine retribution, or as the outcome of a series of very unfortunate events, or as human imperfection screwing up one more thing. Self-destruction is part of our nature, but self-creation is infinitely more powerful. By waking up, metahumans can make right what humans have done wrong. Waking up happens only one person at a time. Reality isn't a numbers game. It's a one-player universe, and you and I are enough to move creation itself.

# A MONTH OF AWAKENING: 31 METAHUMAN LESSONS

O ne purpose of this book is to demystify the process of waking up. The direct path is meant to be effortless and natural. The only uncertainty is time—people who want to achieve higher consciousness all begin at a different place, and this makes a difference. I've found in my own experience that desire is a powerful incentive no matter where you start from. If you really want to learn something—a new language, French cooking, rock climbing—the process becomes enjoyable. The better you get at learning, the more enjoyable it is.

Unlike those things, however, waking up isn't a skill. There's no set of rules or guidelines. Even finding a teacher is fraught with missteps. How does a teacher prove that he's awake? But every culture that believes in higher consciousness has, over time, developed a setting for this special kind of learning, such as an Indian ashram or a Zen Buddhist monastery.

These settings fit into the context of each culture. If you don't happen to belong to the culture, however, and you're viewing these settings from the outside, ashrams and monasteries look alien and exotic. Yet there's no proof that a special setting is mandatory. After all, the process of waking up is about self-awareness. No one can teach you to be self-aware. No one needs to. Consciousness already includes self-awareness. I've been arguing in this book that existence *is* consciousness. In other words, you were born with the tools to become more self-aware. It's just a matter of applying them.

In the process of waking up, no special lifestyle is required. You live as you're living now, by being aware of the world "out there" and the world "in here." The only thing new is that you relate to both worlds using new assumptions. You assume that the "real" reality isn't the same as virtual reality. You assume that you are whole at the level of the true self. You assume that the true self offers a better way to live, a more conscious way but also more creative, open, relaxed, accepting, and free.

Assumptions aren't the same as truth or facts. They need to be tested, which is the purpose of this section. You are asked to experiment on yourself, spending time every day for a month to find out if the direct path works. It's a fortunate time to seek higher consciousness. Stripped of religious trappings and the fog of mystery, consciousness has entered a new phase as a full-fledged research topic, studied by psychologists, psychotherapists, biologists, philosophers, neuroscientists, and even physicists.

This explosion of interest makes for a better setting, in fact, than traditional ashrams and monasteries. You can be fully engaged in everyday life while placing your deeper attention on waking up. It's a little odd that someone must learn to wake up, but that's the result of living so long with the conditioned mind. Convinced that the spell/dream/illusion is real, our minds conform to it. Waking up happens by dismantling the conditioning that keeps us trapped in mental constructs. When those constructs begin to fade away, waking up is the state we arrive at.

Arriving there is unpredictable and totally personal. It's best to set out with an open mind and no expectations. Just adopt the attitude that waking up is real; other people have done it for many centuries, and the only requirement is self-awareness.

# A Daily Plan

The lessons in this section are set up to be as flexible as possible. First there's an axiom or insight for the day. It is followed by a brief explanation, and then an exercise. Read the axiom and explanation at least once, although several times throughout the day is better, so that your attention is drawn back to the theme of the day. The exercise should be done as often as it takes for you to feel that it really sank in—one to three times separated throughout your day should do it. Finally, space has been left for you to journal about your experience. Even better would be to keep a separate journal devoted to waking up.

Is a month enough time to completely wake up? I sincerely doubt it, but some people have been known to open their eyes one morning, look around, and know with certainty that they have awakened. Others gradually change inside and slip into higher consciousness almost without noticing that the change has taken place—it became second nature over the years. In all likelihood you will get the most out of repeating these lessons, going back when you feel the desire to reengage with the learning process. There are degrees of being awake, just as there are skill levels in learning a new language, French cooking, or rock climbing. Reinforcing your awakened state is part of the process.

The lessons get longer as the month goes on, not because they are harder but because there is more to see. Every lesson is equally easy.

Be open, fluid, and flexible about getting there. The beauty of the direct path is that each lesson along the way has its own achievements, its own "aha" moments, and its own pleasures. In that spirit, let the waking up begin.

## DAY 1

*The everyday experience of reality starts with perceptions—*
*sounds, colors, shapes, textures, tastes, and smells.*

Waking up is supposed to be effortless. But it's important to know where to start. There is no better place to begin than where you are right this moment. In fact, we will wind up in trouble if we pretend that there is any other starting point. You are experiencing your life as it is, a flow of experiences that begins with the five senses.

### For Today

Get in touch with the basics. Sit for a moment and be with your simplest experience of light, warmth, the smells wafting your way, the taste of food. Relax into the experience. Just observe. The more you are able to relax, the more effortless waking up will be. Relaxing into the moment is the key. In a relaxed state your mental activity calms down, and observing your direct experience happens naturally.

YOUR EXPERIENCE: _____

_____

_____

_____

## DAY 2

*The range of human perception is a narrow
bandwidth of raw sensations.*

The five senses are our window on reality, but the opening is a slit, not a picture window. "Seeing is believing" usually only applies to a small fraction of the raw data bombarding the eye every second. The same is true for the other four senses. They conspire to deliver a narrow bandwidth of reality. To expand the bandwidth to increase our perception, is one reason for waking up.

*For Today*

Get in touch with how narrow your sense of reality actually is. Cup your hands over your ears and notice how muffled the world is. Put on sunglasses and notice how dim the world becomes. Turn off the lights at night and cautiously, with small steps, try to navigate a room in your house you are very familiar with. When you take your hands away from your ears, remove your sunglasses, and turn on the lights, your awareness of everything around you expands. Waking up expands reality even more.

YOUR EXPERIENCE: _____

_____

_____

_____

# DAY 3

*All biological organisms have their own unique*
*bandwidth of sensory experience.*

Experience defines us all, and since we are tuned in to only one bandwidth of reality—call it the Me Channel—our identity is also narrow. Other living things are tuned in to different bandwidths, giving them an existence we can barely imagine. But humans can change the channel at will. Reality is only as narrow as our awareness. When you wake up, you are tuned in to the entire bandwidth. Then reality is unlimited.

## For Today

Take a moment to listen to the birds singing. Each bird is telling its story. Birdsong communicates information from parent to chick, announces the limits of territory, attracts a mate, signals danger, and identifies which species the bird belongs to. Notice that you do not understand a single thing on Bird Channel. If it is winter or you hear no birds, consider a dog sniffing the air. A dog's nose can tell it who has walked by, what was on that person's shoes, and when the incident occurred. Notice that your nose gathers none of the information that comes across Dog Channel.

YOUR EXPERIENCE: _____
_____
_____
_____

## DAY 4

*Our physical body is also a perceptual experience.*

The Me Channel tells you that you have a body. The body you see and feel, the sensations that come through the nervous system, the locations of pleasure and pain—these signals are constantly broadcast on the Me Channel. The body is not a thing; it is a confederation of perceptions. Your mind unites these fragmented perceptions into a coherent image in time and space. If your mind didn't do that, the Me Channel would just be transmitting noise.

*For Today*

Take a moment to perceive your body directly. Close your eyes and sit quietly. Let your attention roam from sensation to sensation. Lift your arm and feel its weight. Rub your fingers together and feel their softness and the texture of the skin. Hear your breath and your heartbeat. It doesn't matter how many signals you pick up, or whether your body feels nice to you or not nice. You have contacted the real body you have. The experience of the body *is* the body. Everything else is mental interference. When you wake up, you will accept and enjoy the experience of the body for itself, which is blissful.

YOUR EXPERIENCE: _____

_____

_____

_____

## DAY 5

*By itself each perceptual experience is a unique,*
*evanescent, ungraspable, momentary sensation.*
*Our senses take snapshots of reality.*

Life gives us a constant flow of perceptions that we live by. The five senses are the pipeline through which everything flows. But it's not like a continuous flow of water from a faucet. Sensations are much more like rain, which falls one drop at a time. We make sense of life using fleeting thoughts and sensations. We ignore how evanescent every perception really is—each sensation starts to fade as soon as it is noticed. Every thought has already vanished by the time it registers. By waking up, we stop ignoring what is actually happening all the time. The need to turn fleeting sensations into a running movie or story fades away.

*For Today*

Put a grain of salt or sugar on your tongue. Notice how the taste starts to diminish after the first strong taste sensation. Pay attention to how your salivary glands quickly reacted and how your throat wanted to swallow. This puts you in touch with how brief and temporary experience is. But here's the real point. Try to taste what was on your tongue *before* you placed the salt or sugar on it. You can't. That taste, which you probably didn't make note of when it occurred, has fled forever. Fleeting perceptions are the texture of life.

YOUR EXPERIENCE: _____

_____

_____

## DAY 6

*The only constant in every snapshot of perception*
*is the presence of being and awareness.*

Snapshots don't take themselves—there has to be a photographer be-hind the camera. No matter how many thousands of photos a professional photographer takes, he is the constant behind the lens. His job is to look, arrange the setting, place the lights, focus, and decide if the image satis-fies him. You do the same with reality. Your senses deliver snapshots of raw data, which change in endless ways. The only constant is you, seeing, arranging, turning random bursts of perception into something you can relate to. Most of this happens automatically, but when you wake up, you see what you're doing. Then you have much more freedom to create.

### For Today

Get back in touch with the basic units of experience. Sit for a moment and be with your simplest experience of light, warmth, the smells waft-ing your way, the taste in your mouth. Relax into the experience. No-tice each sensation spontaneously, wherever your attention wanders. The more you are able to relax, the more effortless waking up will be. Waking up in itself is a totally relaxed, spontaneous state, open to whatever hap-pens here and now.

YOUR EXPERIENCE: _____

_____

_____

_____

## DAY 7

---

*The stringing together of perceptual snapshots*
*creates a sense of continuity, the same way a movie is*
*created from the rapid sequence of still frames.*

When the invention of motion pictures revealed that our eyes can be fooled by stringing together a series of snapshots at twenty-four frames per second, a deeper truth about reality was also revealed. The human brain works by the firing of neurons. Each firing is a burst of energy, followed by a pause, then the next burst. The bursts slice reality into bits of information from the five senses. When a train races past you, you are not seeing it in motion. You are seeing bursts of information in your brain that give the illusion of motion. Likewise, you don't hear continuous sounds.

The continuity of your life is a necessary illusion. We have to see the world in motion so that we can live in motion, not frozen bits of sensation. Right now you are experiencing pictures and stories created in your mind through the same piecing-together process. When you wake up, these pictures and stories will be seen for what they are: artificial constructs of the mind. You will live from the "real" reality that is beyond pictures and stories—consciousness itself.

### For Today

Sit in front of a moving image on your TV or computer—this can be anything, from people walking around to a news or sports event. Focus on something moving across the screen from left to right. In reality there is no person moving across the screen; not a single photon of light is moving across the screen. Instead, bursts of color are happening, each one totally stationary. By stringing these bursts in a sequence, the illusion of motion is created. Now notice how hard it is to see the actual process taking place before your very eyes. Your mind must see motion because, since you were born, the world has been a series of pictures in motion— this is how conditioned you are to accept an illusion as reality.


YOUR EXPERIENCE: _____

_____

_____

_____

# DAY 8

*The physical body and the appearance of the physical*
*world are created in the mind as constructs from*
*sensations that are intermittent and ephemeral.*

In daily life we do not investigate how the mind creates a three-dimensional world out of random, meaningless bits of sensation. Starting with the simple world a baby experiences, everything gets more and more complicated. A newborn cannot focus on its hand, which looks like a pink blob floating in the air. In time the blob becomes a hand attached to the body; it acquires a name; it develops many skills. Medicine studies it down to every tissue and cell.

This buildup of knowledge occurs in the mind and is created by the mind. A naked hand has no story to tell; it has no developed skills. Everything a hand can become when it belongs to a skilled painter, sculptor, circus performer, chef, or welder is mind-made. The same goes for the whole body and the physical world. We construct virtual reality so that we can have the pictures and stories that are necessary for being human.

*For Today*

Take a sheet of ordinary 8½ × 11-inch paper and poke a pinhole in the center. If you hold the paper close to your eye, you can see the whole room through the pinhole—this is your mental picture of the room. Now hold the paper one or two inches from your eye, until you see only parts of familiar objects—just pieces of lamps, chairs, windows, and so on. Try to walk around the room seeing only these bits and pieces. It is quite difficult. Deprived of the picture your mind makes, the room is a disconnected jumble of fragmented images. Reflect on how you have used the mind to construct the familiar three-dimensional world you accept at face value.

YOUR EXPERIENCE: _____
_____
_____
_____

## DAY 9

*The appearance of the body and the world are*
*activities in consciousness—verbs, not nouns—*
*constantly and rapidly changing.*

When you enter a room, go to work, or take a stroll outdoors, the objects you see appear to be fixed and stable—but they aren't. Your brain is constantly firing to keep the illusion of stability going. Your five senses cooperate by turning photons into pictures and air vibrations into recognizable sounds. In other words, you are constantly making the world. An ever-changing, endless process is taking place in your awareness. Therefore, the outer world is an ever-changing, endless process wearing the disguise of fixed, stable objects. By waking up, you see past the mask of matter, reconnecting with the creative process that makes the world.

*For Today*

Look at a photo of a friend, family member, or celebrity. Now turn it upside down. Notice that you cannot recognize the face anymore. A hitch has occurred in your brain, which is conditioned to recognize faces only right side up. There was a process to recognizing a face; the face itself is meaningless. Or imagine placing a photo on a turntable and setting it spinning (you can try this on a lazy Susan or a record player). Notice that you cannot make sense of the photo while it is revolving. The moving world has no reality until the mind constructs it into a human world. Constant change acquires the illusion of stability and nonchange.

YOUR EXPERIENCE: _____

_____

_____

## DAY 10

*The mental construct of the body and the world
is the product of centuries of conditioning.*

In daily life we accept the world as a given. Trees, mountains, clouds, and sky are simply there. But these are merely the scenery of virtual reality. Everything in the world beyond raw sensory data is rooted in mythology, history, religion, philosophy, culture, economics, and language. Raw sensation is overlaid with this complex conditioning. As a result, the body and the world we perceive have been interpreted in advance. They exist as extensions of the human drama. By waking up, you step out of the drama to be who you really are. You see that virtual reality is a kind of hand-me-down, which you no longer have to settle for.

*For Today*

Here is a simple exercise in perception. Contemplate the letter *A*. When you set eyes on it, you saw a simple sign made with three short strokes of a pen. But those strokes have no inherent meaning, as you can instantly determine by turning *A* on its side or upside down. The meaning of *A* is embedded in it. It is an ancient meaning, going back to the Phoenician alphabet. Mixed in is the Hebrew letter *aleph*, which stands for the beginning, creation, and God. *A* is synonymous with one, which connotes individuality and the start of arithmetic. *A* is a desirable grade in school, and if you earn enough *A*s, you are likely to be well-educated and end up prosperous.

If a single letter of the alphabet carries so much history and so many implications, imagine how complex the fabric of the human world is. We inherit a wealth of meanings that hold the world together but also become a burden. (Think of all the trouble caused by another letter of the alphabet, *I*.)

YOUR EXPERIENCE: _____
_____
_____
_____

## DAY 11

*The mind itself is nothing other than conditioned awareness.*

*Each of us was born into an interpreted world. Previous generations spent their lives giving everything a human meaning.* Every newborn grows up by learning the ropes, and once you can navigate the world—walking, talking, making life choices, forming relationships—you find your place in virtual reality. At some point you would like to have your own unique experiences. "I want to be me" is a powerful incentive.

But the only way to have an experience is to use the mind, and everyone's mind is totally conditioned. It had no choice. By learning the ropes, each of us sacrificed "I want to be me" in the name of "I want to fit in." More than social pressure was at work. The rules of virtual reality require us to accept a shared set of pictures, stories, beliefs, and habits. By waking up, you get to be yourself beyond the rules. The "real" reality is always new and original.

## For Today

Today the challenge is to have a thought that is totally your own. Such a thought cannot echo anything you've heard someone else say or anything you've read in a book. It can't be couched in a familiar phrase. It must not grow out of memory, because then you would only be repeating the past. Faced with this simple challenge, you can see how tightly the conditioned mind holds you in its grip. There are well-tried escape routes, like imagination and fantasy, which get around the rules by not matching reality. There is another escape route, waking up, which allows you to be here now. In the eternal now the conditioned mind has no place.

YOUR EXPERIENCE: _____

_____

_____

_____

## DAY 12

*Virtual reality is a web of relationships.*

The physical world is all about relationships. Around them we create stories. A Christmas tree tells a story; the tree the ornaments are hung on is related to other evergreen trees, which leads back to the plant kingdom and the origins of life. There is nothing in the world that can be seen without being embedded in relationships spreading out in all directions. This web of relationships is the invisible net that holds everything together. Tangled inside the web, we create endless stories in an ongoing movie.

But how do you get out of the web? Humans dream of a realm like Heaven that allows the relative world to drop away forever. Heaven may be a dream, but a world beyond isn't. By waking up, you find yourself in that world, which is consciousness itself. Beyond all created things lies the womb of creation.

### For Today

Let your gaze roam the room and pick any object at random. Now in rapid succession, think of as many words as you can in thirty seconds that relate to the object. Let's say you chose a table lamp. Words related to a table lamp: *light, lightning bug, torch, Statue of Liberty, lamp beside the golden door, freedom, immigrants, Germany, Nazism, Hitler, World War II,* and so on. Notice that the flow of words spreads out on its own, going in any and every direction. By a simple exercise in word association, you have woven one strand of the web that creates the known world.

YOUR EXPERIENCE: _____

_____

_____

# DAY 13

*The mind has entangled us in a virtual reality of our own making.*

In creation myths around the world, God or gods stand apart, looking down on the world they created. For humans, however, we created virtual reality and then stepped into it. The purpose of virtual reality was to allow us a double role, as both the authors of our own stories and the actors who play them out. The two roles are mind-created, and keeping them separate is confusing. When trouble arises, people ask themselves, "Did I do this to myself?" without being able to answer.

Entangled in virtual reality, we find it easier to simply go along and pretend we play only one role, the actor. Yet the role of author is far more important. Unfortunately, how to be an author has been largely forgotten. Life is too confusing already. By waking up, you clearly see your role in the creative process. You are no longer helpless or a victim, any more than Romeo and Juliet are victims of Shakespeare. They came to life in their author's awareness, as you come to life every day in yours.

## For Today

Put yourself back at the creative center of things. The next time you order food or ask to see something in a store, frame the situation this way: *I had the thought that put this situation in motion. I put the thought into words. The words caused another person to undertake a new action. That action prompts another action by the cooks in the kitchen (or the manufacturer who made the goods in a store), who are earning a living to create their own stories, and the sum total of these stories is human history. Therefore, at every moment, my thoughts are at the creative center of history.*

This is more than a new way to frame a commonplace activity. It is the truth. You are the creative center of things, ever and always.

YOUR EXPERIENCE: _____

_____

_____

_____

## DAY 14

*The body, mind, and world, when seen directly and*
*without interpretation, are actually one activity.*

Even though we busy ourselves with a thousand things a day, seeing the world as one thing comes naturally. To a devout believer, the one thing is God's creation. To most scientists, the one thing is the physical universe. But these are conditioned responses. Believers cannot consult God to confirm their belief, and scientists cannot confirm where time, space, matter, and energy come from. What if you look at the world directly, without a conditioned response? You would see that the one thing is awareness constantly modifying itself. Body, mind, and world are experiences in consciousness. This alone can be verified. Experience is the touchstone of reality. When you wake up, it becomes the only touchstone you need. You join the play of consciousness and revel in it.

*For Today*

The play of consciousness embraces all of creation. Today you can join the game as an enjoyable experience. Take a moment to do something that makes you happy—it could be lunch with a friend, appreciating the trees and sky, watching children on the playground. If your enjoyment comes from eating ice cream at midnight, that's perfectly fine. Whatever you are doing, relax into your enjoyment and notice it. Enjoyment is the easiest way to be here now. Just by noticing your enjoyment, you have put yourself into the eternal play of consciousness.

YOUR EXPERIENCE: _____

_____

_____

## DAY 15

*Upon close examination, no external world or physical body can be found independent of our perceptions.*

We are so used to living with a divided self that it is a big step to see beyond it. The divided self tells you that you live in two worlds, one "in here" and one "out there." But if reality is a single thing, this view is mistaken. Consciousness is the one thing. It unfolds as one reality. Knowing this, you have firm ground to stand on—your own awareness. The conditioned mind corrupts and distorts awareness. It colors your perceptions, forcing you to accept the division between the inner and outer worlds. Waking up clarifies the truth. All worlds are experienced in consciousness. There is no need to prove the existence of the physical world or not to prove it. You are here now, and that is enough.

*For Today*

It's not difficult to merge the inner and outer worlds into one. Find a photo of yourself; it can be your driver's license or a snapshot. Holding the photo in your hand, look at yourself in the mirror. Then look at yourself in the photo, and finally, see yourself in your mind's eye. As you moved from seeing your physical body reflected in a mirror, captured on film, and inside your mind, each was an experience in consciousness. On that basis, there were not three different experiences. There was one experience being modified in three ways. Everything in life stands on the same ground, as experiences that are modified consciousness.

YOUR EXPERIENCE: _____

_____

_____

## DAY 16

*Since there is no independent physical world, everyday
reality is a lucid dream taking place in the vivid now.*

Dreams aren't all cut from the same cloth. Some dreams are vague, barely more vivid than having a fleeting memory when you are awake. At the opposite extreme are so-called lucid dreams. When you're having a lucid dream, you have no clue that it is a dream. You are fully immersed in it and, when you wake up, it's hard to acknowledge that the dream wasn't real. Likewise, virtual reality is a full-immersion experience. There are few clues to suggest that you are not fully awake.

For that reason, glimpses of clarity, moments of joy, creative insights, and the experience of meditation are precious. They suggest that you are immersed in a vivid, lucid dream. Waking up will come as a surprise— for many people, it's a shock to realize that they've been asleep all their lives. Every passing moment was like experiencing the vivid now. Once awake, however, now becomes a window into pure consciousness. It doesn't matter what fills the now. What matters is that you are fully awake to it.

### For Today

Moments when you see through the spell/dream/illusion often happen spontaneously—they come upon you by surprise. There's no set way to bring about such glimpses; the closest you can come is through meditation. Still, you can prepare the ground today for the seed of metareality experience. At any time, take a look around, smile to yourself, and say, "Imagine, this is all a dream, and I am the dreamer." The smile is important. It's like anticipating Christmas as a child. You know something good is coming and, by reminding yourself, you open the way.


YOUR EXPERIENCE: _____
_____
_____
_____

## DAY 17

---

*Now isn't a moment in time that can be grasped and*
*held. Now is the rise and fall of awareness.*

If you want to know who enforces the rules of virtual reality, the clock is a good place to start. The tick-tock of clock time slices life into segments of seconds, minutes, and hours. Once you identify with clock time, your life passes in seconds, minutes, and hours. Such an existence is mechanical and routine. Breaking out of virtual reality, now has to become a state of awareness, not a slice of bread. When you wake up, now is a presence; it is the unbroken experience of being here.

Experiencing this presence, you witness how the stream of consciousness delivers a sequence of fleeting sensations and perceptions. Dividing this stream of activity into seconds, minutes, and hours is a mental construct only. When you are awake, you pay more attention to the presence of awareness than to the fleeting events taking place in the mind.

### For Today

Mental activity is very clingy. You have a stake in the thoughts, sensations, images, and feelings that pass through your mind. But you don't have to have a stake in those thoughts, sensations, images, and feelings. Imagine that you are sitting on a commuter train looking out the window. As the scenery rushes by, you don't see it by picking out each building, tree, car, or person. It's all just the passing scenery. If you happen to notice something that stands out, it passes just as quickly as the things you don't notice. Now instead of windows, substitute your eyes. You are sitting behind them watching the passing scenery. When you adopt this position, which is known as "witnessing," you approximate for a moment what the permanent state of being awake is like.


YOUR EXPERIENCE: _____
_____
_____
_____

## DAY 18

*Clock time slices up the timeless, giving it beginnings and endings. As a result, there is birth, aging, and death.*

All of virtual reality, from the atom to the human body to the universe, is a timeless process frozen in time. If you say, "I was born in 1961" or "The meeting starts promptly at three" or "The big bang occurred 13.8 billion years ago," you are doing the same thing—freezing a constant fluid process into a beginning, which automatically brings a middle and an end. Beginning, middle, and end are mental constructs. What is the middle of blue? What was the last thing that happened before time began? When you wake up, being here is continuous—actually, it has always been continuous, until beginning, middle, and end were invented. It will come as a great relief to ditch those concepts. Not only will you find that you are living in the now, but birth, aging, and death will become irrelevant.

*For Today*

To step out of clock time into the timeless, take a moment and look at a color, say the blue of the sky. Try to see beyond the blue. Really try. You will notice that it is futile to try to think your way there. Mental activity is irrelevant. Nor does it matter if you actually see beyond the blue. By stopping your mind from interfering, you escaped clock time, so the only place you can be is timeless. Similarly, try to imagine a time when you didn't exist. This will also stop the thinking mind from interfering. You will experience no time when you didn't exist. Is there a better definition of eternity?

YOUR EXPERIENCE: _____
_____
_____
_____

# DAY 19

*Reality is the endless activity of awareness modifying itself.*

If someone walked up to you and said, "I want to be here now. Where's it happening?" you'd be puzzled. "Now" isn't a place on the map. Brain connections can be mapped in their precise locations, but there is no top, bottom, back, or front to awareness. The now is continuous because awareness is continuous. Only in virtual reality are limitations like beginnings and endings or birth and death imposed. When experienced directly, reality flows like a river. But you have to imagine it as a river that flows in a circle, without starting in the mountains and running to the sea.

When you are awake, even to describe awareness as a flow is too limiting. Awareness doesn't need to be active. As it happens, activity is everywhere. Outside of meditation or unexpected moments of silence, the mind is constantly participating in the rise and fall of consciousness as it modifies itself. Beyond the constant buzz of activity, consciousness is silent, pure, unbounded, not needing to do anything. Upon waking up, you identify with pure consciousness, enjoying the calmness and security it brings.

## For Today

Silent awareness is always with you, waiting to be noticed. Sit in a quiet place and say to yourself, "I am _____," filling in the blank with your full name. With a brief pause between them, say the following things to yourself: "I am [first name]," then "I am," then "Am," and finally no thought. Without labels to identify with, the mind is quiet. As you experience this state, even for a moment, you have found your real identity. The ego emerges from mental activity; your true self emerges from silent awareness.

## YOUR EXPERIENCE: _____

_____

_____

_____

## DAY 20

*Time is only one kind of limitation. So are space, matter,
and energy. Awareness itself is without any limitation.*

Waking up is clearer today than in the past, when the process was
considered so mysterious that it seemed totally paradoxical. As one ancient metaphor put it, wanting to wake up is like being a thirsty fish.
The fish is thirsty only because it doesn't realize that it is surrounded by
the ocean. Likewise, a person seeking to wake up doesn't realize that all
limitations of the mind are the result of not knowing that the infinite
ocean of consciousness is everywhere at every moment.

Limitation begins in the mind but is mirrored by time, space, matter,
and energy. This mirroring effect holds good whether you are asleep or
awake. The difference is that when you wake up, the physical universe is
seen for what it really is, the play of consciousness. Consciousness has no
form or boundaries. Being beyond labels and thoughts, it is inconceivable. It is also who you really are.

*For Today*

Hold up your hand and start moving it into various positions, each
signifying a meaningful gesture. Act out the part of a police officer di-
recting traffic, a teacher pointing to the blackboard, a lover caressing the
beloved's cheek, a chef whipping up an omelet—whatever strikes your
fancy. Reflect on how your hand carried out whatever your imagination
wanted it to. Mind and matter were different appearances of the same
consciousness. Likewise, your personal reality consists of mind acting to
coordinate time, space, matter, and energy. They express the same unlimited possibilities as human imagination.

YOUR EXPERIENCE: _____

_____

_____

_____

## DAY 21

*Virtual reality arose from the human need to live in limitation.*
*This need began the process that created the conditioned mind.*

Limitation is part of virtual reality, and it seems totally convincing and necessary. You cannot fly like a bird; you can't be rich by wishing it were so; if you are hit by a car, you will be seriously injured or killed. I've said that the conditioned mind edits reality so that it serves human needs. The infinite becomes finite. We are hemmed in by harsh realities. There is actually nothing wrong with editing the infinite into the finite— after all, you can't think infinite thoughts all at once, even though you have the capacity to have infinite thoughts.

The problem is that we've forgotten that this editing was done by us. Virtual reality isn't a given; it was manufactured. The setup is finite, and just as you can't think infinite thoughts all at once, humans can't physically do everything at once, say everything at once, or desire everything at once. The setup of virtual reality suits the self we think we have—and must have.

When you wake up, the picture is reversed. You realize that virtual reality is a construct. Only consciousness is a given. Only consciousness cannot be created. By waking up, you free yourself from the conditioned mind, the limited self, and the mind-forged limitations of virtual reality. In freedom you still cannot fly, become rich by wishing it were so, or avoid being injured if a car hits you. On the other hand, it's a bad bet to declare that anything is impossible. Waking up takes you over a threshold. What awaits on the other side is a vast new territory of possibilities.

*For Today*

Sit for a moment and start to think of things that you'd like to do or be that are impossible. You might like to be fabulously wealthy or incredibly attractive or young again—the sky's the limit. As each thing comes to mind, pause and say to yourself, "Why not? Why is this impossible?" Wait for a reply and let it unfold, telling you all the reasons you can't have or be what you want.

Now ask yourself, "Who says I can't?" There is no good answer to this question. Things are impossible not because anybody says so. They are impossible because the whole setup of virtual reality says so. All limitations are built into virtual reality. When anyone says that something is impossible, she is only upholding virtual reality. Who says you have to follow suit? Nobody, including yourself. When this realization sinks in, you begin to glimpse how free you really are.


YOUR EXPERIENCE: _____

_____

_____

_____

# DAY 22

*When infinite awareness is edited, form and
phenomena appear (i.e., things we can see, hear,
touch, taste, smell, and think about).*

For the purposes of everyday life, infinity needed to be edited. Everyone agrees about that, but we forget that there was no rulebook or set of guidelines for editing reality. The only rules are self-imposed. Pure consciousness unfolded into the physical universe, imposing time, space, matter, and energy upon its creation. But pure consciousness didn't believe it had to do things this way. Billions of other universes with different setups have been proposed by modern physics.

Knowing that rules are self-imposed was never in doubt at the level of pure consciousness. We inherited this certainty. Art and culture express the certainty that the human mind can construct any setup while also accepting the setup of the physical universe. This makes it seem that life has two compartments—the mental, which is unlimited, and the physical, which is limited. But that's a mistake. A house isn't separate from the desire to build a house and the knowledge of how it is done. Mental and physical are aspects of one thing: creative intelligence at play. When you wake up, you will see how creative intelligence works, and with fascinated curiosity you will become a co-creator of reality.

## For Today

To become aware of how creative intelligence unfolds, start with a small object. Whether it is a nail, an earring, or the keys to your car, these things are ideas that took physical shape. Now consider something larger, like the Empire State Building or the Golden Gate Bridge. They too are ideas that took physical shape. Does it matter that an earring is tiny and the Golden Gate Bridge immense? No. Creative intelligence isn't large or small. Nor is it hard or soft, here but not there, visible or invisible. The ability to create is entirely self-contained, respecting no limitations of forms and shapes. Creativity needs only itself. Without creative intelligence, forms, shapes, and events couldn't arise.

Now see yourself in the mirror and say to yourself, "I am infinite creativity that has taken on a form." Stop identifying with the form and start identifying with the creativity. This is how you wake up.

YOUR EXPERIENCE: _____ _____ \_\_\_\_
_____
_____
_____

## DAY 23

---

*Every form and phenomenon is actually one thing: the
modification of the formless, the infinite reduced to the finite,
pure awareness given a beginning, a middle, and an end.*

The world rewards big thinkers and gives them a place in history. By
comparison with an Albert Einstein or a Leonardo da Vinci, everyone
feels like a little thinker. But a big thinker isn't necessarily the biggest
thinker. The biggest thinker sees that reality is one thing, creation one
process. That's the whole picture in one view, which becomes your reality
when you wake up.

The whole picture becomes clear when the mind stops constantly
blurring your view with interfering thoughts, sensations, images, and
feelings. They represent the back-and-forth between you and the outside
world. When you wake up, the interference patterns fall away. You accept
as a natural fact that infinite, formless pure consciousness is the source of
all things. What makes this realization natural is that you see yourself as
an expression of one thing, not a jumble of mental and physical activity.

*For Today*

If you see an object in the outside world, you view it from the outside.
Wholeness—the one thing—has no outside. It also has no inside. Therefore, you can't view it. Because you *are* it, you can't even relate to wholeness in various ways, like accepting it or rejecting it, participating one
day and taking time out the next. When you wake up, you know that you
are the one thing. Yet even now you can stop relating to it in false ways.

Today, practice having no attitude toward your mind. Let thoughts
arise and fall, and when you are tempted to have an attitude toward
what is going on, don't. Don't say that one thought is good and another

bad. Don't assign labels like SMART, DUMB, POSITIVE, or NEGATIVE. The mind is none of those things. It is the flow of activity from the absolute. To label the mind is like saying that wholeness is bad, good, positive, negative, and so on. Clearly, wholeness is beyond all labels. So is your awareness. By not judging your thoughts, you begin to adopt the open, nonjudgmental state of being awake.


YOUR EXPERIENCE: _____

_____

_____

_____

# DAY 24

*Only awareness is real. Even as it plays the role of
observer and observed, even as it creates worlds "out
there" and "in here," its own nature is unchanging.*

Awareness exists. Everything else is a passing experience. These two sentences have been said and resaid for centuries. This fact shows, at the very least, that human beings looked at reality and found it mysterious. How did change emerge from nonchange? How did One become Many? The mystery was formulated in countless different ways. Asking "What came before time began?," as modern cosmology does, is merely a variant on the medieval question "What existed before God?"

The answer is contained in the question. What came first has been here forever. Change is just a mask worn by nonchange. Once human beings saw the mystery, we saw our own nature. We are creator and created, One and Many, changing experience and unchanging awareness. None of this has to be proved or tested. Whether you accept your true nature or not has no bearing on your true nature. It continues to be. On waking up, you see your true nature clearly, and then a new life begins.

## For Today

Taking a moment or two, let your mind roam across some things you can recall from your past—memories of early childhood, your parents, birthdays, school, your first kiss, some things that are sad, and so on. It doesn't matter what you choose to see. Now reflect on the one thing they all have in common. You were there. You are the nonchange in the midst of change. Now you know your true nature. Everything else is window dressing.

YOUR EXPERIENCE: _____
_____
_____
_____

## DAY 25

*Human suffering is built into virtual reality.*
*It doesn't exist in awareness itself.*

When you suffer and feel unhappy, anxious, depressed, or hopeless, your suffering feels totally real. You are experiencing something that is a given, feels like a given, rooted in physical pain and mental anguish. But virtual reality, our collective spell/dream/illusion, is a construct. Suffering is embedded in the construct, which is why it feels inevitable. Beliefs about suffering, whether they originate in the doctrine of sin or karma or modern medical theories, reinforce virtual reality.

Waking up doesn't guarantee that there will never again be physical pain or sad days. The conditioned mind is stubborn, and our body has been bombarded constantly by signals from the conditioned mind. Shadows of conditioning continue to fall (always remembering that the shadows can be light as well as heavy). On waking up, you abandon your allegiance to the conditioned mind, and, from that moment, it begins to fade and lose its grip. You see that freedom from suffering is possible and natural. Suffering doesn't exist in awareness itself, which is your true nature.

*For Today*

When people suffer—for example, when they feel depressed or are diagnosed with a life-threatening illness—there's a temptation to blame themselves. "Did I do this to myself?" is a question posed out of guilt, and guilt is quick to blame. The best answer is that your suffering is part of virtual reality. You accepted virtual reality, which makes your participation in pain and suffering inevitable. Not that you are doomed; some people escape serious suffering. But your participation is a sealed deal, no matter who you are or what happens to you, unless you make a change.

To end suffering, break the deal. When you wake up, the whole contract is annulled. Today you can prepare the way by not buying into pain and suffering as something fated and inevitable. Think back on some experiences you consider as suffering—they'd include instances of grief, loss, sickness, betrayal, failure, humiliation, and so on. Now sit quietly, go inward, and be with yourself this minute. You are the same awareness that went through suffering but also not suffering, that experienced pain but also pleasure, that lost but also gained. For every opposite, you have experienced both poles. Therefore, you are not either one.

You are the unchanging awareness that witnesses change, the screen upon which every experience plays without being an experience. This insight holds the entire secret of bringing suffering to an end.

YOUR EXPERIENCE: _____
_____
_____
_____

# DAY 26

*Suffering continues because we cling to memory and
grasp at experience. It is an illusion to believe that the
now can be grasped or that reality can be clung to.*

Truth feels cold when there's nothing you can do about it. People feel
the truth that life brings suffering, and they hate feeling helpless in the
face of this truth. The result is great inner conflict. On the one hand,
we pretend to accept that life brings suffering. On the other hand, we
struggle to escape the feeling of helplessness. Modern medicine ends part
of the confusion and struggle. As diseases are conquered, human beings
feel more powerful and suffering is pushed aside—for the moment.

Suffering in the form of mental anguish has not been alleviated, nor
has the fear of disease and aging, or the dread of death. Upgrading virtual reality is the story of modern technological civilization. (Just as the
discovery of new means of mechanized death is a downgrade of virtual
reality.) Suffering retains its grip because we want to cling to good experiences and the memory of better days. As long as youth, health, and
happiness are rooted in time, with good times being preferable to bad
times, there is no escape from suffering. Clinging to virtual reality means
that suffering is part of the construct.

When you wake up, you don't try to cling or grasp. You don't store up
good memories and push down bad ones. There is only being here now.
In the now there is nothing to hold and grasp, nothing to cling to. By no
longer clinging, you have cut off your connection to virtual reality. Then
suffering no longer clings to you.

*For Today*

When somebody tells you to let go and stop holding on, does that advice really help? The most stubborn resentments, affronts, hurts, and anger are holding on to you, not the other way around. No one wakes up after a bitter divorce, losing a job, or being betrayed by a friend, thinking, "Now I've got something I really want to hold on to." Instead, the anger and resentment come back of their own accord, and they last as long as they decide to last, not as long you want them to last.

What you are actually clinging to isn't bad memories, negative emotions, old grudges, and hurt feelings. You are holding on to virtual reality. By waking up, you let go of your allegiance to it, and then the bad stuff stops clinging to you. Think about something that makes you really angry or resentful. When you have it in mind, let go. You won't be able to, not when it is still clinging to you. In reality, you are where you are. This place is filled with bad things that once happened, hurts and resentments that are in various stages: clinging tightly, starting to let go, or almost faded to nothing. Virtual reality is set up so that, wherever you are, experience clings like barnacles to a ship's hull. Seeing this gives you a sense of detachment, which is a sign that you are waking up.

YOUR EXPERIENCE: _____

_____

_____

_____

# DAY 27

*Suffering comes to an end when we no longer fear impermanence.*
*As long as we have a stake in the illusion, we will suffer.*

Children are eager to explore the world, and they love how everything is changing. But at the same time a child wants the security and safety of home. This balance becomes harder to maintain as an adult. Change becomes threatening when there's no one at home to promise safety and security. One way to hold anxiety at bay is to pretend that "I," the ego-personality, is stable and reliable. "I" has a stake in the world, something to maintain. The ego builds itself up by every kind of clinging—to pleasure, fantasy, wishful thinking, outworn conditioning, old memories, and false beliefs.

All those things are impermanent, so it doesn't work to build a self with them as a foundation. Fear of change departs only when you base your life on the true self. As you wake up, the transition from ego to true self happens naturally.

## For Today

The best way to feel safe is when you take safety for granted without worrying about it. Consider the commercials you see on television for life insurance, pharmaceuticals, retirement homes, and burglar alarms. They offer reassurance by first bringing up the fear that you are not secure and safe. The tactic works because we don't really take our personal security for granted—we push anxiety out of sight instead.

To feel how real, unshakable security feels, stop reading for a second. Then start reading again, and stop again. In the pause between reading these words, you took for granted that you know how to read. There's no underlying anxiety about this—you know it for certain—and the same is true for dozens of things you know how to do. This is how it feels when you are truly secure without hiding from underlying anxiety. When you wake up, you will take for granted that you have always existed and always will. On that basis, you are again like a child, free to explore the world and safe from fear because you are at home always, inside yourself.

YOUR EXPERIENCE: _____

_____

_____

_____

## DAY 28

*Freedom is the natural state of existence,*
*knowing that we are aware here and now.*

Virtual reality isn't reliable when it comes to lasting happiness, security, fulfillment, love, and other things we cherish. Some people enjoy very few of those things, and even when we get more than our fair share, we fear possible loss. It is a bad deal to rely on what is unreliable. You wouldn't take a job from an employer who says that he will toss a coin every day to determine if you stay or get fired. But we cling to virtual reality with no guarantees that things will work out. This is a form of bondage—the worst form, since the widely held belief is that there's no alternative.

Real freedom isn't something you struggle to achieve, hope for, and feel unlikely to win. Freedom is our natural state if we don't imprison ourselves. On waking up, you no longer feel bound to virtual reality. Mental constructs lose their grip and eventually fade away entirely. Being here now is the same as total freedom, because the now is gone before anything can lay claim to you. You exist and you are aware—that's enough to set you free.

## For Today

All of us have our own version of what it feels like to be free and its opposite, what it feels like to be trapped, hemmed in, and suffocated. Yet these concepts disguise the reality, which is that our sense of freedom is always bound up with its opposite. Retirement frees us from the demands of work; having the children go off to college frees us from having them under our roof. But our freedom is constrained by having a job and becoming a parent in the first place.

True freedom isn't bound up with its opposite. To prove this to your-self, think for a moment and describe what you were doing at 7:37 last Tuesday evening. What thoughts were in your head? What words did you say? Even if something memorable sticks with you from last Tuesday, it takes an effort to recall it. You are free from 7:37 last Tuesday evening because there is no attachment to it. A moment of now has fled and gone. The moment when you started to read this lesson has fled and gone. Your relationship to the now is to experience it, extract what it has to give, and move on. This is the state of nondoing and nonclinging, which dawns as your natural state when you wake up.

YOUR EXPERIENCE: _____

_____

_____

_____

# DAY 29

*Knowing ourselves as timeless beings, we can live
consciously. We can be what we really are—a species
of consciousness creating the human universe.*

"I," the ego-personality, grapples with the stuff of creation every day,
turning new ideas into reality. We call this progress, and it is—of a sort.
The ideas we turn into reality have an agenda and a past. They crop up
in a context that accepts or rejects them. The conditioned mind has no
choice but to respond to all kinds of external limitations. Once a wish,
hope, or dream manages to turn into reality, we cling to what we have
created. We ignore the inevitable ruin of all things—one day the things
we've built will be relics like the Parthenon or the Egyptian pyramids.

Creation that is lasting must be built upon the timeless. That's not
possible with physical things, and since physical objects represent ideas,
even ideas cannot truly last in the face of the ravages of time. In the timeless, what lasts isn't an idea or a thing but creativity itself. The "stuff" of
creation is our own awareness and its infinite capacity to create. By waking up, you create on the basis of being a creator, not on the ideas and
things that surround you. You are beyond things and ideas, a conscious
being of timeless dimension.

## For Today

*Timeless* is a concept that feels very far removed from daily life, but
it comes closer when you realize what time is. Time is the process of
creation and destruction. Therefore, if you don't identify with creation
and destruction, you stand in the timeless. You have a choice to shift
your allegiance to the timeless whenever you want. Stop for a moment
to look around the room. When you return to reading this page, every-

thing you looked at is in the process of decaying, dissolving, and fading away. But did time take away the present moment? No—it only took away the things you noticed in the present moment. The present moment constantly renews itself. It stands for the timeless that persists in the activity of time. When you are awake, the timeless takes precedence over everything else. This shift allows you to celebrate creation without feeling anxiety over things passing away.

## YOUR EXPERIENCE: _____

_____

_____

_____

# DAY 30

*Knowing that we are free, the future of humanity can go
beyond birth, death, and all the stories in between.*

If you could land anywhere in history, you could seek out people who
are awake. They would always be a minority, and perhaps a tiny sliver of
a minority. But counting noses isn't how waking up works. If you want
to know if human beings can swim, you only need to find one example.
Likewise, one awakened person tells you that waking up is possible, and,
even more, that everyone is part of the process. Waking up isn't learned;
it is not acquired behavior. It is a state that all of us already exist in. The
only thing that happens when you wake up is that you realize who you
really are. Humanity is already free—we wouldn't be here as conscious
beings if that weren't true. Only the stories we tell ourselves block our
view of our true nature.

This interference pattern, like a fuzzy television picture, doesn't affect
reality. The broadcaster is still sending a clear picture, even if our receiver
isn't picking it up. When you wake up, the signal and the receiver are
both clear and in tune with each other. Thoughts and feelings come and
go without creating interference. This is what it means to be in the world
but not of it.

*For Today*

You can be in poverty for two reasons—either you are actually poor
or you are rich but don't know it. In relation to the infinite potential that
is our true nature, we feel limited in everyday life. So which is it? Are we
actually limited, or do we not know that we are unlimited? The answer
isn't given by looking at conditions on the ground. The richest, smartest,

most gifted, and happiest person can wind up leading a very limited life. The answer is only available in your own awareness.

Sit for a moment and try to think of a forbidden thought. It could be something you have refused to consider for any kind of reason—it is too shameful, outrageous, antisocial, demeaning, or anything else forbidden. The instant such a thought occurs to you, it is no longer forbidden. In fact, no thought has ever been truly forbidden. You cannot limit thought, and since thoughts spring up from silent awareness, you cannot limit the possibility that any thought will be born. Because your entire life—and the life of humanity—is based on consciousness, you too are unlimited. You can stop buying into all the stories about birth, death, and everything in between. Knowing that you are unlimited means that no story can limit your possibilities.

YOUR EXPERIENCE: _____

_____

_____

_____

# DAY 31

---

*You can enjoy the movie while knowing at the same time*
*that you created the movie. That is the awakened state.*

The everyday world operates by opposites. There is a positive and a negative pole to every experience. To navigate through life, people try to grasp the positive pole, but this effort never frees them from the specter that the negative will also have its day. The ultimate polarity is attachment and detachment. Spiritual seekers learn that detachment is positive, because being attached (i.e., stuck, clinging, buying into the illusion) leads to pain and suffering.

In a land where it always rains, it's hard to stop using the word *wet*. In a world ruled by opposites, it's hard to stop using a word like *detachment*. But in the larger picture there is neither attachment nor detachment. Each depends on the other; therefore, each leads to the other. Bad breakups show how hard it is to detach from something (or someone) you were deeply attached to. But we have to relate; that's what the relative world is all about.

In the awakened state, things change. You know that you created the movie, so you can enjoy it without buying into it. This doesn't mean that you detach yourself. Directors love the movies they make. But they also don't keep reminding themselves that they created the movie as they watch it. They take for granted that they are its creator. Likewise, when you are awake you know that you created the movie you are living, but you don't dwell on it. You are too busy being immersed in the now. Being a creator sits at the back of your awareness, and this behind-the-scenes knowledge is enough.

*For Today*

Right now there are things you're attached to and things you aren't attached to. If you are a parent with a young child, for example, you allow your child a certain amount of freedom while also stepping in when needed. This alternation between standing back and getting involved is the day-to-day business of being a parent. Yet in the back of your mind you know that you are a parent; this is your status, and you don't need to bring it to the fore all the time.

Now consider how you parent yourself. In the same fashion, you let yourself go some of the time, while at other times you step in to monitor your behavior. It is impossible to let yourself go all the time and equally impossible to rein yourself in all the time. Yet no matter which mode you are in, in the back of your mind is your sense of self. Sit quietly for a moment and experience your sense of self. Hasn't it always been there, through thick and thin? Your sense of self sits at the back of your mind at all times, not needing to be brought to the fore.

When you wake up, the sense of self looms large at first. You experience with astonishment that everything emerges from your sense of self—all thoughts, words, actions, the outside and inside world. Such a realization can't dawn without bringing a sense of wonder and awe. But in time the sense of self, having realized how infinite it is, retreats once more to the back of your mind. Two people can buy popcorn at the movies, with only one of them a wealthy man. Both eat the same bag of popcorn and pay the same price for it. But the one who knows he is wealthy stores a very different set of possibilities in the back of his mind. This is what it feels like to be awake, knowing that every small action is backed up by infinite possibilities.

YOUR EXPERIENCE: _____

_____

_____

_____

## FOR EVERY DAY

*Waking up is always in the now. There is no time schedule*
*for realizing that you are the dreamer, not the dream.*

A month of awakening has passed. You've taken the journey that goes from here to here. Because there is no distance between here and here, a month was enough to complete the journey. But in another framework, no time is ever enough. Only when the timeless is your playground does "from here to here" no longer matter. The now swallows up all beginnings, middles, and endings.

Being here now was never a goal the way becoming a good person or raising your children right or earning a million dollars is a goal. You cannot compare the now with anything, because every other now is lost forever. What changes as you wake up is subtle but all-important. There is no longer a journey of any kind. Not an outward journey or an inward one. No dream to be fulfilled, no fear to escape from. The past is no longer filled with regrets or the future looming with threats. The phantom side of life has faded away, and those things were phantoms.

*For Every Day*

At random moments, whenever the fancy strikes you, stop and look around. Say to yourself, "A lot is going on. A lot has always been going on. I'm here, and that's what matters." As you wake up, these words will mean something different to you. They will expand to embrace more and more. By being here you are joining the cosmic dance. Appreciate the dance for what it is today. The cosmic part will knock on your door when you are ready.

YOUR EXPERIENCE: _____

_____

_____

_____

# A FINAL WORD

Despite our seven billion stories, we are united in one life. The fate of the planet depends on realizing this fact. Once we do, the human race can evolve to metahuman. How close are we, right at this moment? The answer isn't clear. Life never settles down enough. If a daily newspaper had been available in ancient Rome, medieval France, or Shakespeare's London, the same drama would be seen. The best aspects of human nature always seem precariously balanced against the worst. We are the only creatures capable of self-pity, and the only creatures who feel they deserve it.

Instead of focusing on how humans behave or animals behave or even how quarks and bosons behave, we should ask how consciousness behaves. Consciousness unites everything. All objects, whether a brain cell, a limestone cliff, or a prehistoric flint knife, represent mind in motion. The play of consciousness is infinite, but there is unity holding the play together.

If the dense and challenging ideas in this book leave you baffled, I sympathize. The last thing I want to do—or could do—is to force anyone to accept anything I say. But it is critical to wake up to the one life we share. If you consider yourself a modern person, you live and behave the way the modern world does. As a point of pride, the modern world has been wildly successful at amassing knowledge, one fact at a time. It hasn't been two decades since the number of human genes was counted and the entire human genome mapped. In the coming decade, the brain's trillions of connections will be mapped through a huge, concerted, scientific effort.

So it must be shocking to hear me say in this book that anything you can count, measure, calculate, and reduce to data is part of an all-encompassing illusion. Perhaps even more shocking is my argument that whatever you can perceive, imagine, or think about in words inhabits the same illusion. I am not an enemy of the illusion. I believe that people have a right to upgrade it all they like, and I feel sorrow for people whose portion of the illusion is so degraded that they suffer.

But only waking up allows for the unity of one life to be experienced directly. Otherwise, the world will always be a clash of opposites, founded on the certainty that humans are capable of the best and the worst. What makes us capable of listening to our demons one day and blessing the angels the next day isn't really human nature, though. It is the state of separation we keep reinforcing, generation after generation.

I don't foresee *Homo sapiens* taking a collective leap in its evolution, but I know it can. Our evolution shifted from the physical to the mental domain tens of thousands of years ago, even when early humans were naked and vulnerable, leading an existence as fraught with threats as any animal caught up in the contest of predator and prey. It's a total mystery how we acquired self-awareness. Once we did, or our hominid ancestors did, the mind was poised to triumph in every aspect of life. But the active mind isn't the same as consciousness. A thought, feeling, or sensation is like a wave that rises and falls; consciousness is the ocean.

This analogy goes back thousands of years in India, and I can't remember as a child when I first heard it. The words felt like a cliché,

however, the way that "Love thy neighbor" or "To be, or not to be, / That is the question" feels like a cliché. Repetition leaches meaning away, even from the most profound sayings.

I pondered this obstacle and decided that the direct path has to bring a small awakening every day; waking up shouldn't be held out as the ultimate reward at the end of the spiritual path. In my own life I aim at three kinds of experiences. If one of them happens today, I've achieved a small awakening. If two or all three happen, the small awakening becomes magnified. Here are the three experiences:

I see reality more clearly.

I feel less entangled in habit, memory, outworn beliefs, and old conditioning.

I stop clinging to expectations and external rewards.

How would these experiences apply in your life?

## YOU SEE REALITY MORE CLEARLY

This is the experience of perceiving with fresh eyes. You give up old ways of interpreting the world around you and your own life. Interpretation is built into perception. It is unavoidable that you give names to everything, have opinions, draw on past experiences, and render judgments about whatever is happening. The world has been interpreted for you since you were an infant, and yet you have control over this now that you are an adult. The world doesn't have to change. If you perceive freshness and renewal, if you wake up with a sense of optimism and feel open to the unknown, then every day is a world. You don't have to try to live in the present moment—you won't be able to escape the present moment. It will draw you in without resistance, because there is everything to gain and nothing to lose when a person lives here and now, rather than repeating the past and anticipating the future.

## YOU FEEL LESS ENTANGLED IN HABIT, MEMORY, OUTWORN BELIEFS, AND OLD CONDITIONING

This is the feeling of getting unstuck. Virtual reality would be perfectly acceptable if people felt free to alter it according to their own desires. But a great deal of life is beyond our control, which leads to a feeling of being trapped, confined, limited, and even suffocated. I lump these feelings into the phrase "getting stuck." By getting unstuck you disentangle yourself from the complicated web woven by the ego-personality. This web has become sticky through the process of identification. Anytime you say "I am X," you are further away from being able to say "I am." As we saw, X can be anything: your name, job, marital status, race, religion, nationality. These and much more become your personal story. "I am" is beyond all stories.

## YOU STOP CLINGING TO EXPECTATIONS AND EXTERNAL REWARDS

This is the experience of becoming your true self. The ego-personality is constantly on the lookout for external rewards to validate its worth. If you ask people, "Which would you rather be—happy simply to exist or rich?" their answer is obvious. The need for external rewards, not just money but status, the right neighborhood, a new car, social approval, and more, fuels our dependence on them. Over time the ego-personality has become a dominant force, even when someone considers himself unambitious or spiritual—the front pew in church feels better than the last row. But the ego is a false guide, because the total fulfillment it promises is always over the horizon. Living on expectation goes hand in hand with needing external rewards—there is always a mythical jolt of pleasure or triumph or wealth, the big score that will make life worthwhile once and for all. To cut the strings, you need to feel that the absence of external rewards isn't painful, because it is offset by inner rewards. The greatest of these is the freedom to be yourself.

As a result of these three experiences, waking up becomes your life, little by little, and then, almost without knowing it, you are participating in the one life, which is real, radiant, and whole.

Because everyone can have small awakenings, the future of our species doesn't have to be a grand project marked by the great upheavals of war and peace, revolution and backsliding, achieving greatness and losing it again, or playing the roles of oppressor and oppressed. One person at a time can awaken to reality. It will be enough. The mystery of being human has been hidden from each of us, which may be why it remains so tantalizing. We have never stopped being a self-created species. If we can create a world of glorious highs without being fully awake, imagine what we could do with our eyes wide open.

# ACKNOWLEDGMENTS

I feel gratitude every time a new book is finished for the generous and productive relationship I have with my publisher. Beginning with Gina Centrello, president and publisher of Penguin Random House, who has shown unwavering loyalty to me. Thank you.

At Harmony Books, I enjoy a trusting relationship with my perceptive editor, Gary Jansen, who is untiring in offering suggestions to improve the manuscript. I can't count how often Gary has guided my writing in the right direction—an author can hardly ask for more. In addition, I'd like to acknowledge the support of everyone at Harmony Books who dedicated their time, creativity, and passion to this book, including Aaron Wehner, Diana Baroni, Tammy Blake, Christina Foxley, Molly Breitbart, Marysarah Quinn, Patricia Shaw, Jessie Bright, Sarah Horgan, Heather Williamson, Kellyann Cronin, and Ashley Hong. And a special word of thanks to Rachel Berkowitz, in the foreign rights department, who has been instrumental in helping to spread my work around the world. The public has little idea of how dedicated the team at a publishing house needs to be and how much they love books and serve writers. Many thanks.

A special thanks to Poonacha Machaiah, an inspired innovator, good friend, and wise guide.

Finally, there is everyone closer to home. I would like to thank the teams in California and in New York: Paulette Cole, Marc Nadeau, Teana David, Sara McDonald, Aaron Marion, Angie Lile, Attila Ambrus, and Kendall Mar-Horstman.

The newest collaborators in my life are the Infinite Potential podcast team: Jan Cohen, David Shadrack Smith, Julie Magruder, and friends at Cadence 13—thank you for allowing me to be connected to the digital world along such a productive new path. While *Metahuman* was being written, a year of new challenges faced the Chopra Foundation board of directors, who adapted with fantastic support and guidance, so deepest thanks to Alice Walton, Matthew Harris, Ray Chambers, Francois Ferre, Fred Matser, Paul Johnson, and Ajay Gupta, along with the "explorers" who are on the journey of awareness. I'd also like to welcome Tonia O'Connor, CEO of Chopra Global.

My family has grown over the years, gone through many changes, and yet continues to be a source of loving warmth and joy: Rita, Mallika, Sumant, Gotham, Candice, Krishan, Tara, Leela, and Geeta, I hold you in my heart forever.

# INDEX

*The ABC of Relativity* (Russell), 121
acquired savant syndrome, 81–82
*Advaita*, 175
*ahimsa* (harmlessness), 12
alien intelligence, 117–118, 119
Allen, Woody, 208–209, 211
Antarctica, 68
anxiety, 24, 124–125, 153, 324–325
apoptosis, 128
arctic fox, 71
Aristotle, 39
artificial intelligence (AI), 23, 205–206
art of not-doing, 253, 254–256, 262
associative memory, 265
Astin, John, 24, 25n
authenticity, 24, 217, 224–225, 226
awareness, 140–141, 216–217, 308–309. *See also* body, freedom from; choiceless awareness; consciousness; pure consciousness; self-awareness
AWARE study, 38–40
Ayurveda lifestyle, 31, 32

bees, 264–265
big bang theory, 112–113, 115–116, 121–122, 162, 227
Billé, Cardinal, 274
biorhythms, 127–129, 221
birds:
    adaptions by, 68, 71
    migration of, 127–128
    songs of, 284
    talking to, 11–12
Black Death, 107–109
Blake, William, 16, 49, 59, 224–226
Blanke, Olaf, 41
bliss, 45–46, 186–187, 250–252
body, freedom from:
    anatomy of awareness and, 216–217
    exercises for, 217–222
    lessons on, 285
    modified consciousness and, 130–132
    timelessness and, 224–230
    wholeness and, 215–216, 222–224
body temperature, 202–203
Book of Genesis, 115–116
bottlenecks (genetic), 87–88, 98

boundaries, 51–52, 99, 122–123, 191,
    202, 219
brain:
    anti-robot argument, 19–21
    consciousness without brain
        function, 39
    development of, 141–143, 146, 148,
        150, 200
    fail-safes of, 125
    freeing your body from, 216,
        220–221, 223–224, 228–229,
        234–235
    going beyond, 5–6. See also
        metahuman
    hormones of, 3–4
    mental model theory and, 41–43
    mind versus, 93–95, 141, 143–145,
        146, 204–205, 210, 267
    "other-brained" people, 80–82, 84
    reducing valve of, 148–150,
        151–153
    time and, 126
    vagus nerve and, 220–221
    virtual reality's effect on, 36, 242–
        243, 288–289, 292
Buddha, 106, 178, 239–240, 241
Buddhism, 106, 186, 254
"Burnt Norton" (Eliot), 119
Butlein, David, 24, 25n
butterflies, 71, 117–118

Caesar, Julius, 103
Capra, Fritjof, 118
cave paintings, 90–92
cells:
    communication between, 54–55
    differentiation of, 199–200
    programmed death of, 128
cerebrum, 141–142
Chalmers, David, 109
Chauvet-Pont-d'Arc Cave paintings,
    90–92
cheetahs, 73–74
chimpanzees, 65, 87–89

Chit Akash, 219
choiceless awareness:
    about, 253–254
    art of not-doing, 254–256
    direct path to, 249–254
    exercise for, 256–257
    practical immortality and,
        258–262
Chopra, Deepak:
    Life After Death, 37n
    Quantum Healing, 269
    Super Genes (with Tanzi), 21
Chopra Center study, 31–32
Christianity, 106, 254, 258, 272–273
circadian rhythm, 127
cognitive dissonance tolerance, 24
cold-blooded animals, 203
Cold War, 67
collective consciousness, 126, 263–
        267, 300, 337
color, 51, 166–167, 168
compassion, 24, 145
congenital savant syndrome, 81–82
consciousness. See also body, freedom
        from; creativity; self-awareness;
        waking up; specific types of
        consciousness:
    about, 2–6, 17–18, 57–58
    analogy for, 105, 338–339
    choosing, 6, 11–12, 22–23
    as creation experiencing itself, 120
    defined, 105
    evaluating levels of, 24, 30–32
    evolution of, 69–72, 106–107,
        273–274
    existence and, 141, 267
    human evolution and, 101–102,
        103–104
    as infinite, 74–75, 77–80, 148–149,
        312–313
    metareality and, 22–23, 94–95,
        114, 250
    as mind source, 105, 143–145,
        146–147

misconceptions about, 196–198
multiple dimensions of, 89–92
out-of-body experiences and, 37–40
phobias and, 124–125
physical expression of, 79–80
physiological basis for, 40, 93–95
rationality and, 111–114
reason and, 106–111
research on, 31–32, 38–40, 183–191
self and, 170–175, 178–179, 203–207, 211–212
as self-creator, 138–139
stories and, 106–107, 116–118, 119
timelessness and, 224–230
as uncreated, 140–141
as universal, 118–119
virtual reality and, 35–37, 111–114, 160, 162
coral reefs, 201–202
cosmic consciousness (supreme enlightenment), 113, 267, 268–272, 275
creation stories, 115–119, 137–139, 298, 318–319
creativity:
lessons on, 298
modified consciousness and, 129–136
potential for, 84–86
practical immortality and, 260
self-creation as, 137–140
traits of, 92

Dante, 273
default mode network (DMN), 149–150, 151–153
defensiveness, lack of, 24
Dennett, Daniel, 110
depression, 97–98, 151–152, 153
Descartes, René, 141
de Waal, Frans, 88

direct path:
about, 193–198, 246–247
to choiceless awareness, 249–254
overview, 250–254
practical immortality and, 258–259, 262
self-regulation and, 201–204
sense of self and, 203–212, 241–247
"so what?" test and, 207–212
to waking up, 194–198, 206–212
wholeness of universe and, 198–200
divided self, 98–99, 102–103, 105
*Divine Comedy* (Dante), 273
divine perfection, 273
DNA. *See* genes
dogs, 3–4, 65, 259–260, 265–266
dreams, 132, 133–134, 302–303

ecstatic state, 85–86
ego:
about, 59–61
agenda of, 61–64, 72, 74–75, 81, 180, 185, 197
current state of, 67
default mode network (DMN) and, 150, 152, 153
desires of, 172, 174
evolution of, 64–67, 70–72
identification with, 171, 173–174, 178
lessons on, 324–325, 328–329
transitioning from, 73–75, 173–174, 211–212, 233, 309, 324, 328, 340–341
Einstein, Albert, 50, 94, 114, 121, 129–130, 269–270
Eliot, T.S., 119, 126, 171–172
emotional discomfort tolerance, 24
emotions, 19, 71, 185
enlightenment:
consciousness and, 106–111

enlightenment *(cont'd)*:
  cosmic consciousness and, 267,
    268–270
  as expanded self-awareness, 191
  methods for, 187–191
  stages of, 183–187
*Enlightenment Now* (Pinker), 107–110
everyday world. *See* personal reality
evolution:
  of art, 91–92
  of consciousness, 69–72, 106–107,
    273–274
  of ego, 64–67, 70–72
  of humans, 67–72, 87–89, 101–102,
    103–104, 139
  metareality embedded in, 86–89,
    93
  of self-awareness, 105, 338
  of sense of self, 67–69
  survival of the fittest, 134–135
existence. *See also* choiceless
    awareness:
  art of not-doing and, 255
  as consciousness, 141, 267
  lessons on, 318–319, 326–327
  as uncreated, 140–141
experiences:
  of enlightenment, 183–187
  explanation of, 160–163
  lessons on, 295, 300
  quantum domain and, 163–168
  reification of, 168–170
  root of, 160
  true self and, 170–175. *See also* true
    self
"eyes open, no thoughts," 234–235,
  237–238

fail-safes, 126–129
Feynman, Richard, 130, 271
flow, as "in the flow," 89–90
freedom, of thought, 204–205
Freud, Sigmund, 153, 170, 241,
  249–250

Garden of Eden, 105, 119, 273
genes (DNA):
  anticipation by, 128–129
  development and, 199–200
  evolution and, 87–89, 199
  metareality and, 88–89
  misconceptions about, 19–21,
    53–58
giant anteater, 86–87
giant panda, 87
Google (workplace), 85–86
gorillas, 65, 87, 88, 89
gratitude, 24, 264
great apes, 65, 66, 88
Great Barrier Reef, 201–202

Hacking Creativity project, 86
*Hamlet* (Shakespeare), 97–98
happiness, 90, 190–191, 249–250,
  251–252, 254
Harari, Yuval Noah, 22–23
Harvey, William, 107–108
Hawking, Stephen, 113
hearing, 163–165
Heaven, 250–251
Heisenberg, Werner, 18, 268, 269
higher consciousness, 6, 11–13, 31–
  32, 183–191, 279–281
Hinduism, 258
"The Hollow Men" (Eliot), 171–172
*Homo Deus* (Harari), 22–23
Host levitation, 274–275
*How to Change Your Mind* (Pollan),
  149, 151–152
Hoyle, Fred, 115
human evolution, 67–72, 87–89, 101–
  102, 103–104
human nature:
  about, 272
  choiceless awareness and, 249–254
  duality of, 98–99, 236–237,
    249–250, 320–321, 322–323,
    337–338
  lessons on, 320–321, 322–323

liberation from, 11, 24, 184–185,
   225–230, 240–241, 272–275
self and, 170–175
human potential:
   about, 77–80
   brain and, 93–95
   multiple dimensions of, 82–89, 98,
      100–105, 148
   sudden genius phenomenon and,
      80–82
humility, 24
Huxley, Aldous, 148, 149, 150,
   153–154

Iceman, 100–102
illusions. *See* virtual reality
insight, 12, 85, 244–245, 259
instinct, 70, 127–128, 147, 235–236
"in the flow," 89–90
intuition, 145, 171, 244
Iry-Hor (pharaoh), 64

Johnson-Laird, Philip, 41
*Juramaia*, 139–140

Kabir, 253
Kant, Immanuel, 108
Kierkegaard, Søren, 254
Koch, Christoph, 264–265
Kotler, Steven, 47–48, 84–86
Krishnamurti, J., 49–50, 238

language, development of, 143, 145,
   146–147
Lascaux (France) cave paintings, 90
Lemaître, Georges, 115
Leonardo da Vinci, 177–178
leptin, 20
Levandowski, Anthony, 205
levitation, of Host, 274–275
life, blueprint for, 53–58
*Life After Death* (Chopra), 37n
light, going into, 38–39
Lin, Tao, 150–151

Linde, Andrej, 111–113
Lourdes (France), 274
*Love's Executioner* (Yalom), 122–124
"The Love Song of J. Alfred
      Prufrock" (Eliot), 126
lucid dreams, 302–303

*The Magic Flute* (Mozart), 255
Marean, Curtis W., 68–69
Martin, Jeffery, 183–191
Maslow, Abraham, 10–11, 13
Maya, 269
McKenna, Terence, 151
meditation, 44–49, 85–86, 154, 189–
      190, 235, 238–239
Menon, Krishna, 174–175
mental model theory, 41–43, 49
metahuman:
   about, 1–5
   consciousness-is-the-ocean
      analogy for, 105, 338–339
   defined, 11, 13, 98
   as freedom, 98
   mental constructs as barriers to,
      208
   organized innocence of, 49–50
   on sense of self, 60–61
   shift to, 5–7, 337–339. *See also*
      metahuman lessons; waking up
   traits, 24, 30–32
metahuman lessons:
   about, 279–280
   daily plan, 281
   for days 1 to 31, 282–334
   for everyday, 335–336
   further lessons for every day,
      335–336
   research on, 31–32, 187–191
metareality:
   about, 5–6, 22, 58
   consciousness and, 22–23, 94–95,
      114, 250
   as database for human experiences,
      83–84, 86

metareality *(cont'd)*:
    evolution and, 86–89, 93
    infinite potential of, 77–80, 312–
        313, 314–315, 332–333
    lessons on, 306–309, 312–315,
        320–327, 332–333
    metahuman survey, 23–32
Metzinger, Thomas, 41–42
micro-dosing, of psychedelics,
    152–153
mind. *See also* pure consciousness:
    about, 229
    brain versus, 93–95, 141, 143–145,
        146, 204–205, 210, 267
    consciousness and, 105, 143–145,
        146–147
    defined, 105
    development of, 147–148
    intuition and, 145
    learning and, 15
    lessons on, 295
    psychedelic use and, 149–155
    reification by, 168–170
    sense of self and, 238–239
    source of, 143–144, 146–147
    traits of, 260–261
Mind at Large, 148, 150, 153,
    154–155
mind-body connection:
    about, 20–21, 215–216
    anatomy of, 216–217
    exercises for, 217–222
    interpretation shift and, 51–54
    personal reality and, 20–21
    physiological basis for, 3–4
    research on, 20–21, 31–32,
        47–50
    timelessness and, 224–230
    transformation of, 222–224
    wholeness and, 222–224
mind-reading (theory of mind),
    88
Minsky, Marvin, 122
miracles, 274–275

mirrors:
    invention of, 64–65
    personal recognition and, 51, 52,
        58, 59
    toddlers and, 59
monarch butterflies, 71
monism, 252. *See also* choiceless
    awareness
Montgomery, Sy, 266–267
mortality, 258–259
Mozart, Wolfgang Amadeus,
    255
multiverse, 269

*Nature: Human Behavior* (journal), on
    gene testing, 20
*Nautilus* (journal), on DNA as code
    of life, 54
near-death experiences, 37–40, 37n,
    151, 250
Nelson, Adrian David, 113
NETI (Nondual Embodiment
    Thematic Inventory)
    questionnaire:
    about, 23–24
    questions, 25–30
    score evaluation, 30–32
    study on, 31–32
    therapist scores, 30
*The New Yorker*, on out-of-body
    experiences, 40–42
nonviolence, 106
not-doing, art of, 254–256

obesity gene, 20
*Observer* (newspaper), on
    consciousness, 17
observer effect, 167–168
octopus, 263–264, 265–267
One-Electron Universe, 271
one life:
    causeless cause, 272–275
    collective consciousness, 126, 263–
        267, 300, 337

cosmic consciousness, 113, 267,
    268–272, 275
  surfing the universe exercise,
    270–272
organized innocence, 226
Ötzi "Iceman," 100–102
out-of-body experiences, 37–43, 151
oxytocin, 3–4

pain:
  enlightenment and, 47–49, 187
  Menon on, 174
  subjective experience of, 47
painted lady butterflies, 117–118
Parnia, Sam, 38–39
participatory universe, 272
peak experience, 11–13
Peck, M. Scott, 181–182
penguins, 71
perceptions, 282–289, 292–294, 339
Perfect Health (Ayurveda-based
    program), 31–32
persistent non-symbolic experience,
    184
personal reality:
  about, 9–13, 310
  anti-robot argument and, 19–21
  as house of illusions, 17–18
  limitations of, 73–75
  phobias and, 134–135
  questionnaire on, 23–32
  sense of self and, 59–60
  virtual reality and, 21–22, 36–37,
    59–61, 125
phobias, 124–125
Pinker, Steven, 107–110
Pinnacle Point (South Africa), 69
placenta, 140
Planck, Max, 17
Plato, 39
Plato's forms, 268
Pollan, Michael, 149, 151–152,
    154
postcreation, 272

practical immortality, 258–262
precreation, 272
prodigies, 145
Proust, Marcel, 265
psychedelic use, 149–155
psychoanalysis, 249–250
psychology graduate students, 52
psychotherapists, 30
pure consciousness:
  about, 231–233, 251–252
  cosmic consciousness and, 268
  creation story and, 273
  desire, to remove desire, 237–241
  exercises for, 233–237
  lessons on, 308–309, 310, 314–315,
    316–317
  sense of self and, 241–247

quantum domain, 163–168
*Quantum Healing* (Chopra), 269
quantum revolution, 17–18, 36,
    56–57, 78–79, 111–114, 115,
    269–272
quantum vacuum, 53, 138, 227, 232,
    259

*The Radiance Sutras* (Roche), 46
rationality, 108–109
reality, infinite. *See* metareality
  reality, limited. *See* virtual reality
Reality Sandwich, 188
reason, 106–111
red knot (sandpiper), 127–128
reification, 168–170
relativity, 121–122
religions. *See also specific religions:*
  on creation, 272–273
  promises of, 193–195
  rejection of, 178
  rules of immortality, 258
resilience, 24
Richardson, Ken, 54–55
Richie, Michael, 151
RNA, 54–55, 199

*The Road Less Traveled* (Peck), 181–182
Roche, Lorin, 44–46
Rose, Carl, 233
Rothman, Joshua, 40–42
Rumi, 194, 208, 239
Russell, Bertrand, 121

sandpipers, 127–128
savant syndrome, 81–82
*Scientific American:*
  on collective consciousness, 264–265
  on human survival, 68–69
  on sudden genius phenomenon, 81
sculptures, 65–66
seabirds, 71
self-awareness. *See also* pure consciousness:
  anatomy of, 216–217
  consciousness and, 280
  defined, 105
  embracing self-awareness, 6–7, 140–141, 190–191, 211–212, 215–216, 279–280. *See also* metahuman lessons
  evolution of, 105, 338
  of fail-safes, 125
  of metareality, 88–89, 95
  physiological component of, 149–154
  as reality disruption, 4–6
  reduction of, 148–149
  sense of self and, 206–207, 211–212
  true self and, 178–179
self-creating universe, 137–138
self-creation, 137–139
self-pity, 337
self-regulation, 201–204
sense of self:
  about, 209–212
  as divided self, 99, 102–105

evolution of, 67–69
getting stuck and, 340
lessons on, 284, 324–325, 332–333
metahuman on, 60–61
personal reality and, 59–60
self-awareness and, 206–207, 211–212
versions of, 170–175
virtual reality and, 59–61
wholeness of the mind and, 238–239
senses:
  illusions of, 51–52
  lessons on, 283, 286, 287
  physiological basis for, 14–15
  quantum level of, 163–168
  reification of, 168–169
  virtual reality perception by, 21–22, 36, 50
Shakespeare, 97–98
Siegel, Mikey, 47–48
sight, 163–166, 167–168
silent awareness, 308–309, 330–331
simulated reality. *See* virtual reality
smell, 163–165
snow petrel, 68
solipsism, 175
*The Soul of an Octopus* (Montgomery), 266–267
"so what?" test, 207–212
speech, 142–143
Sri Atmananda, 174–175
*Stealing Fire* (Kotler and Wheal), 47–48, 84–86
stories:
  consciousness and, 106–107, 116–118, 119
  creation, 115–119, 137–139, 298, 318–319
  postcreation, 272
  transcendence blocked by, 179–183, 235–237
stress reduction, 31, 185, 220–221

struggle, absence of. *See* choiceless
        awareness
sudden genius phenomenon, 80–82,
        83, 190
sudden savant syndrome, 81–82
Sufis, 261
*Super Genes* (Chopra and Tanzi), 21
superior autobiographical memory,
        190
superstitions, 107–108
supreme enlightenment (cosmic
        consciousness), 113, 267, 268–
        272, 275
surrender, propensity to, 24
symbols, 261
synaptic pruning, 142
synchronicity, 256

Tagore, Rabindranath, 1, 2
Tanzi, Rudy, 21
Taoist teaching, 254
*The Tao of Physics* (Capra), 118
taste, 164–166, 168, 286
theory of mind (mind-reading), 88
therapists, NETI Questionnaire
        scores for, 30
Thomas Aquinas, 272
time:
        consciousness of, 126–129
        experiences of, 227–229
        lessons on, 304–305, 306
        modified consciousness of,
                129–132
timelessness, 224–230, 304–305, 306,
        328–329
touch, 164–165
transcendence. *See also* enlightenment:
        about, 176–178
        blocking, by clinging to stories,
                179–183, 235–237
        methods for, 187–191
        states of awareness and, 183–187
Treffert, Darold, 80–82

*Trip* (Lin), 150–151
true self:
        about, 5, 170, 172–175, 178–179
        aging and, 223
        engaging with, 242, 280
        "so what?" test for, 207
        transitioning to, 235, 238, 324,
                340
truth, interest in, 24

uncertainty principle, 167
unconscious self, 171–172, 173–174,
        175
universe:
        accepted mental constructs of,
                120–125
        creation of, 115–119, 137–141
        fail-safes for, 126–129
        modified consciousness and,
                129–136
        surfing exercise, 270–272
        wholeness of, 198–200, 201, 203
Upanishad, 2

vagal breathing, 220–221
Venus of Berekhat Ram, 66–67
Venus of Tan-Tan, 66–67
violence, 102–103, 106
virtual reality:
        about, 13–16, 35–37
        blueprint for, 53–58
        brain, effect on, 36, 242–243, 288–
                289, 292
        as clash of opposites, 338
        consciousness and, 35–37, 111–114,
                160, 162
        construction of, 162–163
        editing of, 69–74, 77–78
        ego's agenda and, 61–64
        experience and, 160
        fail-safe for, 120–125
        going beyond, 44–50, 77–78
        illusion of, 17–18

virtual reality *(cont'd)*:
  interpretation shift and, 51–54
  lessons on, 288–300, 312–313,
    320–323, 326–327
  letting go of, 73–75. *See also* virtual
    reality, freedom from
  Menon on, 174–175
  mental model theory on, 41–43,
    44, 49
  out-of-body experiences, 37–43
  personal reality and, 21–22, 36–37,
    59–61, 125
  reification and, 168–170
  sense of self and, 59–61
  senses on, 21–22, 36, 50
  "so what?" test for, 207–212
  as three-dimensional symbol,
    261–262
virtual reality, freedom from:
  contradictions of, 102–105
  defined, 97, 98
  divided self and, 98–99, 102–103,
    105
  as enlightenment, 184–185
  enlightenment and, 103–105,
    106–111
  lessons on, 326–327
  letting go, 73–75
  overview, 97–99
  quantum physics and, 111–114
  violence and, 97–98, 100–102

waking up:
  daily plan for, 281
  direct path to, 194–198, 206–212
  embracing self-awareness, 6–7,
    190–191, 279–280
  lessons on, 300, 301, 302–303,
    304–305, 330–331

process of, 159–160, 279–280, 281.
    *See also* metahuman lessons
  research on, 187–191
  self-regulation and, 201–204
  sense of self and, 203–207,
    209–212
  small awakenings, 339–341
  "so what?" test for, 207–212
  stages of, 185–186
  to wholeness of the universe, 198–
    200, 201, 203
warm-blooded animals, 203
Watson, James, 55
Way of the Future (church), 205
Wheal, Jamie, 47–48, 84–86
Wheeler, John, 271, 272
White, E. B., 233
whole mind, recovering. *See* pure
    consciousness
wholeness:
  immortality of, 259, 262
  lessons on, 316–317
  of mind-body connection,
    222–224
  resting in, 219–220
  as universe, 198–200, 201, 203
Wilde, Oscar, 223
*Wired* (magazine), on artificial
    intelligence, 205
witchcraft, 107–108
witnessing, 246
wonder, 49–50
Wordsworth, William, 62
*Wu wei*, 254

Yalom, Irvin, 122–124
yoga, 154–155, 220, 243

*Zelig* (Allen), 208–209, 211

# ABOUT THE AUTHOR

....................................

DEEPAK CHOPRA, M.D., FACP, founder of the Chopra Foundation and cofounder of the Chopra Center for Wellbeing and Jiyo, is a world-renowned pioneer in integrative medicine and personal transformation, and he is board certified in internal medicine, endocrinology, and metabolism. He is a Fellow of the American College of Physicians; a clinical professor in the Department of Family Medicine and Public Health at the University of California, San Diego; a researcher of neurology and psychiatry at Massachusetts General Hospital; and a member of the American Association of Clinical Endocrinologists. The *WorldPost* and *The Huffington Post* global Internet survey ranked Dr. Chopra as #17 of the most influential thinkers in the world and #1 in medicine. Chopra is the author of more than 85 books translated into over 43 languages, including numerous *New York Times* best-sellers. *Time* magazine has described Dr. Chopra as "one of the top 100 heroes and icons of the century."